KNIFE
SKILLS

HOW TO CARVE / CHOP
SLICE / FILLET

D0452512

KNIFE SKILLS

HOW TO CARVE / CHOP SLICE / FILLET

MARCUS WAREING • SHAUN HILL
CHARLIE TROTTER • LYN HALL

WITHDRAWN

DK

LONDON, NEW YORK,
MELBOURNE, MUNICH AND DELHI

Project Editor Annelise Evans
Project Art Editor Phil Gamble
Senior Editor Jennifer Latham
Senior Art Editor Isabel de Cordova
Managing Editor Dawn Henderson
Managing Art Editor Susan Downing
Production Editor Jenny Woodcock
Production Controller Sarah Sherlock
Creative Technical Support Sonia Charbonnier
Senior Creative Nicola Powling
Assistant Jacket Designer Rosie Levine

First published in Great Britain in 2008 by
Dorling Kindersley Limited
80 Strand, London WC2R 0RL
Penguin Group (UK)
This paperback edition published in 2012

Copyright © 2008, 2012 Dorling Kindersley Limited
Text copyright © 2008, 2012 Lyn Hall, Shaun Hill, Charlie Trotter,
Marcus Wareing. For further details see page 224.

Some material in this publication was previously published
by Dorling Kindersley in *The Cook's Book*, 2005

4 6 8 10 9 7 5 3
008-KD117-Aug/12
All rights reserved. No part of this book may be reproduced, stored
in a retrieval system, or transmitted in any form or by any means,
electronic, mechanical, photocopying, recording or otherwise,
without the previous prior permission of the copyright owner.

A CIP catalogue record is available from the British Library

ISBN 978-1-4093-7664-4

Colour reproduction by MDP, UK
Printed and bound in China by Hung Hing

Discover more at
www.dk.com

Learning Services		
Cornwall College St Austell		
Class	6433 WAR	
Barcode	ST026731	
Date	02/15	Centre	B

CONTENTS

PART 1

KNIFE SKILLS: the basics

CUTTING-EDGE TECHNOLOGY 8

Learning to cut; anatomy of the knife;
history of knife-making; modern
knives; choosing a knife; types of knife;
cutting tools; cutting accessories;
garnishing tools; knife safety; storing
knives; caring for knives

PART 2

KNIFE SKILLS: the application

VEGETABLES 50

Peeling, slicing, dicing, coring, stoning, shredding; cleaning vegetables; making ribbons, crisps, cups, French fries, chips; preparing stir-fry; tomato concassé; chilli flowers; chopping herbs; preparing spices

FISH & SHELLFISH 98

Round fish & flat fish – gutting, boning, trimming, skinning, filleting, serving; slicing raw round fish; shellfish – oysters, clams, scallops, prawns, langoustines, crab, lobster, squid, octopus

MEAT, POULTRY & GAME 138

Beef, lamb, pork & ham – boning, slicing, butterflying, carving; offal; chicken – jointing, spatchcocking, boning, cutting up cooked birds, carving; other birds – turkey, duck, goose; jointing a rabbit

GLOSSARY 214

RESOURCES 216

SAFETY & FIRST AID 217

INDEX 218

ACKNOWLEDGMENTS 224

DOUGHS & DESSERTS 182

Raw doughs – cutting & stamping doughs and pastries; finishing doughs; baked doughs – filling pastries & cakes, making Melba toast, slicing sandwiches, pastry & cakes; chopping & working with chocolate

FRUIT 194

Apples – peeling, slicing, fluting; sliced pear; lemon zest julienne; orange sections; pineapple – slices, chunks, fluted; peeling peaches & nectarines; preparing mango; melon boats & balls; Macédoine salad; chestnuts

PART 1

KNIFE SKILLS: the basics

CUTTING-EDGE TECHNOLOGY

By learning to use a knife properly, not only will your presentation of food improve and it appear more appetizing, but in many cases the food will taste better. For example, take a butter sauce, Beurre Nantaise – many homemade versions become sour and acid because the shallots are not cut finely enough to release their sugars.

You will also find that choosing the best-quality knives that you can afford will turn out to be a blue-chip investment: you cannot lose, because you will find preparing and cooking foods of all kinds much easier and quicker.

choosing and using knives

As you will see in the following pages, there are considerable differences in kitchen knives, for example in the types and grades of steel – from corrosive and rigid to stainless and elastic. Over the centuries, knife blades and other tools have been developed so that modern chefs have at their disposal a variety of cutting edges to suit different foods and preparation methods, as well as to provide for the needs of different cuisines.

Once you have acquired a suitable collection of knives and blades, this book will show you how to use them correctly and skilfully. The professional chefs spend years perfecting their knife skills and pride themselves on mastering the exactness and precision that can come only with practice. By following the advice and illustrated steps in this book, you too can learn how to choose a knife so that its handle fits your hand like a mate, how to hold it intimately so that it becomes an extension of your arm and hand, and how to control its action.

For example, you will learn how to use blades with round-edged points to rock, slide, and chop vegetables effortlessly or to select a more aggressive, long, scalloped knife to bite evenly through a crusty sour-dough loaf.

While using knives, remember that you are responsible for your own safety and that of everyone else in the kitchen, but do not let this deter you from the thrill of wielding a precision tool with skill and efficiency.

maintaining knives

Do not misuse a knife: it is not a can-opener or a screwdriver. You will irreparably damage your knife and could easily injure yourself. There is little more dreary in life than struggling with a blunt knife, so keep it sharpened and honed – this cannot be done too frequently. A ceramic or Japanese knife, when sharpened and honed daily by professionals, can "drop" through an onion, without any juice escaping.

LEARNING TO CUT

Many people are hesitant about using knives with dexterity or assume that such skills are unattainable. Nothing is further from the truth.

Learn how to wield three sharp knives and you will halve the time it takes to prepare a meal and make time in the kitchen much more satisfying.

Three common faults prevent people from cutting properly. The first is trying to carve a cold cut by pressing the knife through the meat. Unless used on butter, knives cut only when in motion and sliding backwards and forwards like a saw. The second mistake is to make a pert, rocking movement of the blade, as if the food needed help to fall off the blade. The third error is choosing too small a knife for the task.

Before you start, make sure that the work surface is the correct height for what you are going to do. Straightforward cutting and carving usually is done on a 90cm (3ft) high work top, butchery and work with a cleaver on a lower surface. Check that the chopping board does not make the work surface too high up in relation to your body. Your cutting board is a place of action, not a storage area for peelings or cut items. Before you begin, set out a couple of small baking trays and a bowl for waste, so you can clear and clean your board constantly.

Begin with the messiest tasks first, such as plucking pheasants and snipping pinions, gutting fish, or simply peeling vegetables. As you progress to finer cutting and decorative tasks, you will be able to concentrate on and enjoy them, knowing that the big stuff is done.

correct ways of cutting

First, adopt a well-balanced and relaxed stance: place your feet hip-width apart so your weight is carried by the centre of your feet. Relax your shoulders and arms for optimum flexibility in your wrists. A long knife is best for slicing and carving tender ingredients – some chefs peel an orange with a 36cm (14in) scalloped slicing knife. A long blade and flexible wrist gives more precision than trying to remove peel with a tightly gripped paring knife. Most pastries are cut with long knives for this reason. Think of violinists and the control that they have over their bows on the strings: their wrists are raised above their fingers, and their forearms driven by the elbows and shoulders. The more you slice a knife to and fro, with minimum pressure, the better you will cut.

Try carving one breast of a well-cooked chicken with a long knife, while flexing your wrist as fast as you can. Then try forcing the same knife through the other breast. The first chicken breast will yield

chopping vegetables **Cut vegetables according to how they are cooked: from large chunks for slow cooking and sauces to fine dice for releasing flavour into oils.**

moist, juicy, and elegant slices. The second, although just as tender, will appear dry, ragged, fragmented, and overcooked.

The rocking-chopping action is a comfortable way of chopping and renders consistently thin slices. Place the ingredient flat on the board and parallel to the edge of the work surface, and use a large chef's or santoku knife. The blade tip never leaves the board, helping to bring the cutting edge closer to the last slice. Drop your wrist so that the heel of the knife slides down on the item, cutting it, and push the knife tip away from you, through the ingredient, towards the other side of the board. Raise your wrist and the knife, point still on the board, and bring the heel a whisper away from the previous cut, to produce the thinnest possible slice. Once you have mastered this technique, it will be a real pleasure to reduce mounds of ingredients to thin slices.

For soft vegetables and fruits, like apples and mushrooms, resort to the mandolin for fast and precise results with no practice whatsoever.

ANATOMY OF A KNIFE

A good knife is a highly effective tool, perfected through centuries. Each part of the knife has properties designed to deal with the wide range of textures and tasks found in the kitchen. Understanding the role that each part plays is crucial to using a knife correctly. One knife cannot efficiently cut every ingredient. For example, the most important parts of a large knife are the bolster and the heel (which is ideal for heavy chopping and cleaving), while a small knife with a fine point and tip to the blade is the best tool for more intricate work.

the point
is used to make fine incisions, and to pierce foods, papers, and films (eg cling film before microwaving)

the spine is the top of the blade, is wider in large knives, and may be grasped by the fingers for better stability; in some knives, it is also useful for crushing garlic

the tip (the first third of the blade) is used for cutting soft vegetables, small ingredients, through ligaments, and for fine slicing

the cutting edge, between heel and tip, works hard during chopping and slicing

Knives differ significantly in shape and size, even within the Western world – the flexibility of a long blade in a fish filleting knife and the curve of the short blade in a turning knife enables both to perform specific tasks in the kitchen. However, the basic structure remains the same.

At first glance it might be difficult to distinguish a high-quality knife from an inferior one, but if you look at the knife shown below, you cannot help but admire its lean lines. From its incisive point to the well-finished head of the handle, you can see that it is fit for its purpose.

the **bolster** is the junction between the blade and the handle, and protects the hand in large knives

the **tang** is not always visible, but is the part of the steel that extends into the handle

the hollow **rivets** hold the tang to the blade; they should be flush and tight with the handle, with no crevices for bacterial growth

the **handle** may be made from a variety of materials and is important to comfort in cutting

the **heel** is the heaviest part of a large knife and closest to the hand; it is used with maximum strength to cleave through hard, tough foods

a classic forged chef's knife
All knives, irrespective of size, have a similar anatomy.

types of kitchen-knife blade

Your knife collection should feature a variety of blades. This helps to keep them all sharp: if you use scalloped blades when appropriate, those with tapered ground blades will get less wear. You will also enjoy cutting all ingredients, precisely and without bruising, in many different recipes.

tapered ground edge
Seen in cross-section, this blade has a thin, long, taper from spine to cutting edge. Most knives have this blade: it is good for chopping and general-purpose slicing.

scalloped edge Often on a long knife, the teeth of this edge are similar to those in a serrated blade. The teeth protect the cutting edge and the blade stays sharp for longer. Ideal for cutting fragile sponge cakes (*p189*).

serrated edge The saw-toothed edge easily bites through tough skins and tender centres, as in aubergines. If only one edge of the blade is serrated, eg some tomato knives, you can sharpen it at home (*p44*).

granton edge The oval depressions hold tiny pockets of air, which stop slices that are being cut from sticking to the blade. This blade is good for moist foods such as large joints of cold meats, smoked salmon, ham, and turkey.

single ground edge Japanese knives were traditionally bevelled on the right-hand side only. They were thought to produce cleaner slices.

types of kitchen-knife handle

Handles vary a great deal and have a big impact on your enjoyment of using a knife. When buying a knife, think of the tasks you need it for and assess the handle accordingly. A large knife should have a big handle and all knives should feel comfortable in your hand.

wooden handle **Formerly, these were made of hardwood, eg rosewood or walnut, with a tight grain that needed occasional oiling. Today, wooden handles are composite and impregnated with plastic.**

plastic handle **Plastic handles are smooth, crevice-free, and now most popular. Polypropylene (shown here) handles are cheaper, can be slippery, and may melt near hot pans. Handles (see below) of polyoxymethylene (POM) give a better grip and last longer.**

ergonomic handle **The flowing lines are designed to fit the contours of your hand and to be extremely comfortable. Many chefs use a knife for up to 16 hours a day, so a good grip is essential.**

"rat tail" tang

the tang

Cheap knives have a "rat tail" tang, which is mean and thin – a millimetre wide, running just 3cm (1½in) into the handle – or have no tang at all. A full tang, found in the best knives, indicates that the blade runs the entire length and width of the knife, giving excellent balance, stability, and endurance. Sealed plastic handles hide the tang, otherwise it is visible on both edges of the handle and is fastened in place by rivets.

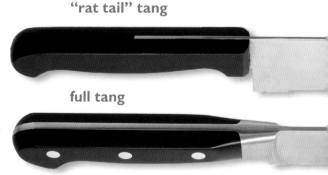

full tang

HISTORY OF KNIFE-MAKING

Man's ingenuity has produced cutting tools for millions of years – first for use with stone and then with food. Today, knives and scissors have been designed and developed for every purpose in the kitchen. Magnificent professional knives, made from stainless steel with a high carbon content and used by the world's leading chefs, are readily available for everyone.

early cutting implements

Early cutting implements were made from stone, ivory, horns and antlers, but by 6500BC humans had discovered how to mine and extract the metals copper, lead, and gold. These were too soft for hunting and cooking implements – even blending them with other metals and minerals to produce alloys, such as bronze, did not solve the problem.

By 4000BC, the Egyptians were using knives made from obsidian (a polished, volcanic glass) and flint, which gives a good cutting edge. The real boost to knife-making was the discovery of iron, around 1000BC. Iron bestowed strength and durability for cutting and chopping. It was also cheap and available for common use, but was prone to rusting and also too malleable.

With the mastery of smelting around 700BC, metalsmiths added carbon to iron to make steel, reducing the danger and difficulty that had marred earlier attempts. Improvement of the furnaces allowed more control to produce a metal that was durable, flexible, and able to take and hold a sharp edge.

knife-making in the West

Kitchen knives developed in small forges out of the production of side weaponry such as daggers, sabres, and swords. In the fourteenth century, Chaucer mentions a cutler in Sheffield ("cutler" was the name then given to a maker of knives and weaponry) and the town is still a British centre for knife-making.

modern western knives **A knife begins as a steel blank (far left), a long strip of metal – high-quality steel is used for the best knives. The blank is hammered out on a forge to delineate the rough shape of the blade and tang on the blank (centre left). The shape is cut out of the blank and holes drilled in the tang to take the rivets that will attach the handle (near left).**

modern knife manufacture

machine grinding **During manufacture, both sides of the the blade are ground repeatedly to a fine edge; water is used to cool the blade and prevent any deterioration. The blade is then hardened in an oven.**

checking by hand **The blade is hammered and straightened by hand and checked repeatedly during and after the process. Then the tang is straightened in a vice before a handle is attached.**

By the sixteenth century, the French were making the finest knives in the world; René Antoine Ferchault de Réaumur wrote a treatise on metallurgy in 1722. Table knives, spoons, and forks had become part of European culture. However, carbon steel proved to be too soft, was easily pitted, and discoloured by acidic foods, and the cutlery required careful and immediate drying. By 1912, however, greater control of the furnaces became possible and stainless steel was produced by adding chrome to carbon steel. This new steel didn't rust or discolour and produced a tough blade with a sharp edge, which was hard to attain, but once produced held in wet conditions.

By now the Germans were the master cutlers of the Western world. In 1731, in Solingen, the powerhouse of knife-making, Peter Henckels had registered the TWIN trademark with the Solingens Cutlers' Guild. His company mixed carbon steel, iron, chrome, and other metals to make high-carbon stainless-steel knives, with a superb cutting edge.

knife-making in the Far East

Among all the exciting, distinctive cuisines of Asia, the Chinese and the Japanese deserve especial recognition. Eating small morsels with chopsticks demands expert cutting and chopping to have taken place beforehand. The standard knife in a Chinese kitchen is a large, carbon-steel, square-ended cleaver, and it has been so for centuries, although now they are available in polished stainless steel. In contrast, the range

of Japanese knives – hand-sharpened to *honbatsuki* ("true edge")
standard – is legendary. There are two types of traditional Japanese
knife: *kasumi* and *honyaki*.

kasumi & honyaki knives

These knives derive from traditional Samurai sword manufacture.
Making a kasumi knife involves a complex process of heating high-
carbon steel and soft iron together, hammering the alloy flat, folding
it, then hammering it flat and folding it again. This hand-working of
the two metals is repeated, in many layers, and often at various angles.

When the blade is polished, a shimmering but subtle pattern is
created – called *kasuminagashi*, the "floating mist". It is also known as
the Damascene effect after the laminating process, which evolved in
Damascus, Syria, after 400BC. From AD1300, Sakai became the capital
of small weaponry manufacture in Japan. Knife production started in
the sixteenth century, when the Portuguese introduced tobacco to
Japan, and knives were needed for cutting it.

Honyaki knives are of higher quality, being made entirely of high-
carbon steel, but they are more difficult to use and to maintain their
kirenaga, or duration of sharpness.

the knife-making craft in Japan

During the Genroku period (1688–1704), the very first *deba hocho*
knives for cutting vegetables were produced: knives with curved spines
and lethal points, with the arched grace of a ballet dancer's pointed
toe. The knives' extreme sharpness allows food to be cut into the
thinnest of slices without ragging.

This was followed by a wide range of kitchen-knife styles, all with
traditional handles of *honoki* wood, from a species of magnolia that was
also used by sword makers. Blades varied from extremely long
and thin, used to cut tuna, to blunt-ended cleavers. The Tokugawa
shogunate (1603–1868) granted a special seal of approval to the Sakai
knife industry, which virtually gave it a monopoly.

Miki City is a centre for traditional blacksmiths and silversmiths.
Most knife manufacturers are still small family businesses, where
craftsmanship exceeds volume and they produce only a few knives
a day. Seki City is considered the home of Japanese kitchen cutlery.
Technology has updated ancient forging skills to produce world-class
stainless- and laminated-steel knives. In the *san mai* (a three-layered,
laminated blade), metal layers are laid evenly, like a baker making puff
pastry, which results in a blade that resists corrosion and maintains
strength and durability. Handles are often made of hardwood.

The blades are much longer and the tips more pointed than Western knives. Japanese knives are used with great precision.

MODERN KNIVES

Thanks to rapid advances in technology, both Western and Japanese knife manufacturers produce a huge range of knives to a very high quality. It is up to individuals to choose knives of different properties that suit the food and recipes prepared in their own kitchens.

Japanese knives

In Japanese culture, the preparation and presentation of food are raised from routine, daily tasks to an art form. Japanese knives are central to this tradition and their evolution has been driven solely by functional requirements. The blade, with a hard, brittle core that takes and holds a supreme edge, is supported and contained by ductile metal cheeks that protect the core, leaving just the cutting edge exposed. This gives the knife great strength and durability.

These knives require more care than Western kitchen knives: they should be washed by hand, sharpened frequently using Japanese waterstones, and occasionally wiped over with a light mineral oil. Japanese chefs will do this every night after service – a task that is more a religion than a duty. In return, they have knives with superb balance and sharpness that make food preparation a true pleasure.

Traditionally Japanese knives are bevelled on one side only, for use with the right hand. Although requiring more skill to use, it was thought that this would give a cleaner cut, and would be easier to maintain the sharpness against a stone.

fusion knives

Recently, traditional Japanese knives have been transformed to meet the demands of the Western kitchen. The new knives are made with

Japanese knife

rounded **tip** allows rocking, slicing, and chopping

traditional **handle** of magnolia (*honoki*) wood

single right **bevel** stops sticking and allows food to slide up and out double-bevelled blades from very pure, stain-resistant, high-carbon steel that is alloyed, like Western knives, with molybdenum. This is a transition metal, used in high-strength steel alloys, that does not react with oxygen or water at room temperature. The result is an extremely sharp and hard edge, but it is still cushioned and protected in the traditional way by softer, outer layers of chrome stainless steel, so that only the cutting edge of the core is exposed. The santoku knife (p27), similar to a large chef's knife, was designed for the West and adopted with gusto in the Western home kitchen.

Ceramic blades are also becoming more popular, in spite of their higher cost. The key ingredient of this type of blade is zirconium oxide, which produces a blade of incredible sharpness that lasts for years if cared for correctly. The ceramic blades can be sharpened only by professionals, on a diamond wheel, and may snap if handled carelessly.

Western knives

The best of the Western knives, although forged, are not laminated or worked in the same way as in Japan. There is only one way of making Western knives: high-quality steel is hammered out on a forge and then the blades are sharpened to a fine cutting edge by grinding on both sides (p17).

Recent refinements include freezing blades to below –70C° (–94°F) for better protection against rust – even stainless steel is prone to rust. They are also heated twice to just below 300°C (572°F), to harmonize the molecular structure of the metal.

Western knives are sharpened from time to time on a stone, but honed throughout the day on a steel. Annually, each knife should be treated professionally on a grinding wheel, to keep the blade straight and trim, otherwise it will wear in a curve. Blades with serrated and granton edges provide extra scope for cutting but, again, will need professional sharpening.

greater **heft** copes with
larger amount of food efficiently

Western knife

shaped **handle**
for comfort and safety

large **blade** used as a
cleaver and for heavy chopping

CHOOSING A KNIFE

Take plenty of time in choosing a new kitchen knife. A good-quality knife is expensive, and should last a lifetime, so give this moment the consideration it deserves. Buying a knife at a demonstration provides an ideal opportunity for asking pertinent questions of an expert, gaining specialist advice, and – most importantly – giving the knife a test run.

how many cheap knives do you have?

As with many tools, cheap knives are not cost-effective and it is always worth investing in as good a set of knives as you can afford. Quality knives are either Swiss, German, or Japanese. These countries have long been leaders in the manufacture of precision instruments, having access to steel foundries, and a history of uncompromising perfection.

If this is your first knife, choose a chef's knife (p24), which will suit most tasks. When loosely grasped, the handle should fill your hand – you won't want to put it down. Run your thumb around the curve of the bolster; look for a substantial blade of forged steel, with a pleasing satin finish. If you hook your index finger around the bolster, the knife should swing gently into a horizontal orientation. If choosing a slicing or filleting knife, press the tip against the counter: it should flex in an

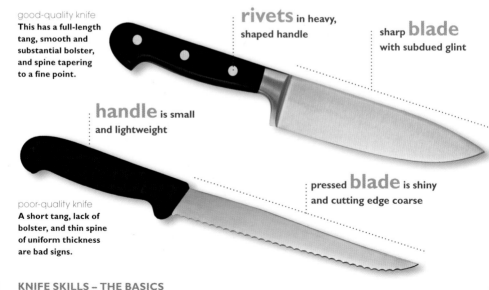

good-quality knife
This has a full-length tang, smooth and substantial bolster, and spine tapering to a fine point.

rivets in heavy, shaped handle

sharp **blade** with subdued glint

handle is small and lightweight

pressed blade is shiny and cutting edge coarse

poor-quality knife
A short tang, lack of bolster, and thin spine of uniform thickness are bad signs.

well-balanced knife **The full tang gives greater heft to the handle, distributing the weight evenly through the knife, to the point of the blade, giving excellent balance and making the knife easier to use.**

arc, with buoyant resistance. Big knives should be strong and heavy near the bolster and heel of the blade, to help you when you chop by cleaving powerfully through large and tough ingredients.

The blade should feel cool and contain a good proportion of carbon, with enough chrome to keep it bright. The higher the amount of carbon, the easier it is to keep sharp. Razor-sharp blades cut easily, without pressing or mashing the ingredients. If you nick a tomato, the flesh should spring back instantly.

A new blade should bite and cling to a wooden surface and be wide enough to scoop up finely chopped ingredients, such as parsley. The spine should be distinct, full. and strong near the handle, and taper to a fine wafer at the sharp point. If possible, test its performance and how it feels in your hand by trying it out at cutting and chopping.

a knife for life

There is no doubt that using properly sharp knives involves an element of danger, and that they should be handled with care and respect. Remember that you will use a good-quality knife every day. It will enhance your experience of cooking by making the preparation of the ingredients much easier and a real pleasure. So if you don't get a thrill from handling a particular knife, the knife is not for you.

using the **correct**
knife is the **first step**
to cutting like an
expert chef

3 *utility knife*

2 *filleting knife*

1 *boning knife*

electric knife

An electric knife is not essential, and
should never be used to carve a roast,
but comes into its own with dealing
with delicate, finished items that have
taken a lot of work, such as a terrine
(*p158*), a puff pastry tart, or a cake
(*p192*) with a nut and caramel
topping. It cuts superbly clean slices.

TYPES OF KNIFE

Although knives with plastic or stainless-steel handles can be cleaned in a dishwasher, it is best to clean them by hand. In a dishwasher, stainless steel can rust, wooden handles swell, and it is extremely dangerous to stack knives in a dishwasher pullout.

1 **boning knife** This is ideal for taking the flesh off bones and silver skin off meat, such as silverside and fillet. A slightly flexible blade works best. The blade is short, to give full control of the tip as it works at 180° angles, delving deep into joints and snipping ligaments. Slide it parallel and along the curve of the bone; never cut on to the bone, which would blunt it. 2 **filleting knife** Essential if you cook fish, this is the most flexible blade of all. When pressed hard against the backbones of flat fish and employed in flat, sweeping or "wiping" movements, this knife takes fillets off the bones. It deftly cuts round fish like mackerel or trout in half lengthways. 3 **utility knife** Usually under 15cm (6in), this has a finer blade than the chef's knife, and may not have a full tang. It is good for carving and slicing fine, white meats such as chicken breasts or calves' liver, and soft vegetables like courgettes. 4 **paring knife** With its sharp point and virtually no bolster, this is essential for all the fiddly jobs with small fruits and vegetables, such as coring out seeds of chillies. It is also good for peeling or slicing items held in the hand, like carrots and apples. 5 **chef's knife or cook's knife** This general, all-purpose kitchen knife gains its versatility from the size of its blade, which ranges between 15cm (6in) and 36cm (14in). The largest chef's knives can crack open crab claws and coconuts, kill lobsters, and chop huge bunches of parsley, as well as cut hard, tight cabbages. The long blade is useful for cutting raw doughs and pastry. The shorter blade is best for smaller vegetables, for slicing and trimming fruit and vegetables, eg oranges, and working with meat and fish. Both sizes of knife can be used for cutting meat, mincing and dicing herbs and vegetables, and for julienne.

4 paring knife

5 large chef's knife

6 **slicing knife** The long, elegant blade of this knife ranges from 18–26cm (7½–10¼in) in length and is easy to manipulate precisely. It is perfect for slicing boned meat and slides over the breastbones of game and poultry, carving the tender fillets. The point lifts out the juiciest bits of meat from next to the bone. 7 **15cm (6in) serrated knife** Sometimes called the "gin-and-tonic knife", this handy knife will save wearing out your other knives. The teeth bite easily through tough skins and safely cut through round ingredients that tend to roll. Use it for saucisson, tomatoes, and lemons. 8 **scalloped slicer** At 28cm (11in), this is ideal for crusty French and Italian breads and large fruits, such as melons. It is also excellent for slicing fancy, iced cakes; dip it in hot water between each slice to clean the blade. 9 **granton knife** The 30cm (12in) blade is designed to carve and cut even thin slices. Held horizontally or vertically, its length gives great control and precision. The indentations hold air, ensuring that slices do not stick together, so is ideal for cheese, cured salmon, cold cuts, and pâtés. 10 **bread knife** Essential in every kitchen, a bread knife cuts through crusts and enables you to keep knives with tapered ground blades perfectly sharp. Avoid bread knives with pressed blades. A good bread knife will cut precisely through mounds of sandwiches with chunky fillings, breads for croûtes, and pain bagnas.

11 **santoku knife** Inspired by Japanese knives, the "stir-fry knife" has a blade that is good for scooping up finely sliced vegetables and herbs. It is easy to learn to chop with and the rounded end is safe in the hand of a beginner. The knife makes fast work of making uniform vegetable dice, so they cook quickly but stay crunchy, and is a good stand-in for a cleaver.

12 **turning knife** This knife is for cutting large, often misshapen vegetables, like potatoes or cucumbers, into useful, perfectly neat, little rugby balls. It is also good for showing off!

slicing knife

15cm (6in) serrated knife

scalloped slicer

bread knife 10

santoku knife 11

granton knife 9

turning knife 12

16 palette knife

17 trowel spatula

18 dough cutter

15 Chinese cleaver

14 fish scaler

13 clam knife

20 oyster knife

13 **clam knife** The rounded blade is slightly sharp on one side and designed, like the oyster knife, for easing the mollusc off the shell. Recent models have a plump, round handle for a better grip, and are made of non-slip plastic, so can be used dry or wet. There is no guard on this knife, since smooth clam shells pose little threat.

14 **fish scaler** This tool has teeth to rip off the scales. The handle must feel comfortable when gripped tightly, for exerting pressure. The scaler is always used under running water to rinse off the scales and should not rust. 15 **Chinese cleaver** A traditional, weighty blade of razor-sharp carbon steel and wooden handle helps with chopping and slicing all foods. Cooks in the West use it for heavy work; deft Chinese chefs employ it for everything, even jobs like deveining prawns, and the exigent demands of wok cooking. 16 **palette knife** At about 30cm (12in) long and beautifully balanced, this knife has many uses. It lifts and turns long, delicate fillets of cooked fish without breaking them, acts as a guide for cutting dough or pastry, and is vital for lifting rolled-out pastry that has stuck to the work surface.

17 **trowel spatula** The dog's leg on this ultra-efficient version of a palette knife lets you slip the blade easily under food. It is ideal for sweeping melted chocolate evenly over a cake, pressing crumbs and a melted butter crust into a cheese-cake tin, and smoothing a roulade into a shallow oven sheet. 18 **dough cutter** The blade is uniformly flat and efficient, but not sharp, and may be straight or rounded, with a wooden or stainless-steel handle. It is used for chopping risen dough to make rolls or buns. Straight-bladed cutters serve also as a valuable scoop for chopped nuts, herbs, and spices, and for clearing a work surface of fine debris. 19 **scalpel** The surface of a fully risen dough is as fragile as a butterfly's wing. To score it attractively, as in the loaves of French baguettes, without dragging and irretrievably spoiling it, you need a scalpel. 20 **oyster knife** A traditional handle is thick, solid, and made of hardwood, and is large enough to allow a good grip when thrusting, then twisting the small, sturdy blade to open a shell. A large guard protects your fingers from the rugged shells as you force the blade between them. The blade is also fairly blunt, to ease the live oyster off the shell. 21 **Parmesan knife** Not so much a sharp-edged cutting knife, this short-bladed knife will dig into a large round of matured Parmesan cheese. Since the cheese has a granular nature (it is not a pressed cheese), it will break into large, appetizing slivers or nuggets, for eating or grating.

19 *scalpel*

Parmesan knife

21

SHARP-BLADED CUTTING TOOLS

There are dozens of tasks that a professional chef does easily with a knife, but that we find difficult – reach instead for a pair of kitchen scissors or shears. The mandolin and mezzaluna, too, will enable you to perform in the kitchen like a professional chef.

3 mezzaluna

1 **kitchen scissors** Use cheaper scissors for paper, flowers, wire, and cardboard, to keep your kitchen scissors sharp. They should have non-slip handles (for use dry or wet), pointed tips to snip ligaments in the centres of tight joints, and sharp blades to cut herbs or soft bones, eg of trout or quail. Teeth on the blade are used in loosening bottle tops. 2 **poultry shears** The long curved blades and longer handles cut powerfully through ribcages and backbones of game birds, and trim leg bones of game. Use to cut through, rather than carving, joints and for de-stemming large bunches of lovage and parsley. 3 **mezzaluna** The curved blade rocks fast, safely, and efficiently to chop soft and hard ingredients, such as herbs, chocolate, and nuts. Large handles give a comfortable grip. Double-bladed versions are harder to clean safely, but deal with more ingredients at a time.

4 **mandolin** This gives a short cut to enviable slicing, including julienne, grating, and ribbons. It stands on firm legs at a perfect, ski-slope angle for sliding ingredients over the adjustable blades. Using the carriage is advisable for beginners and vital for making chips. Stand with feet shoulder-width apart, weight in the centre of your body, and hold its legs firmly against your middle with one arm and hand. Slide each vegetable or fruit over the blade carefully; as slices fall away, flex your fingertips upwards. The faster you cut, the better the cutting action, and the finer the slices.

kitchen scissors

1

poultry shears

2

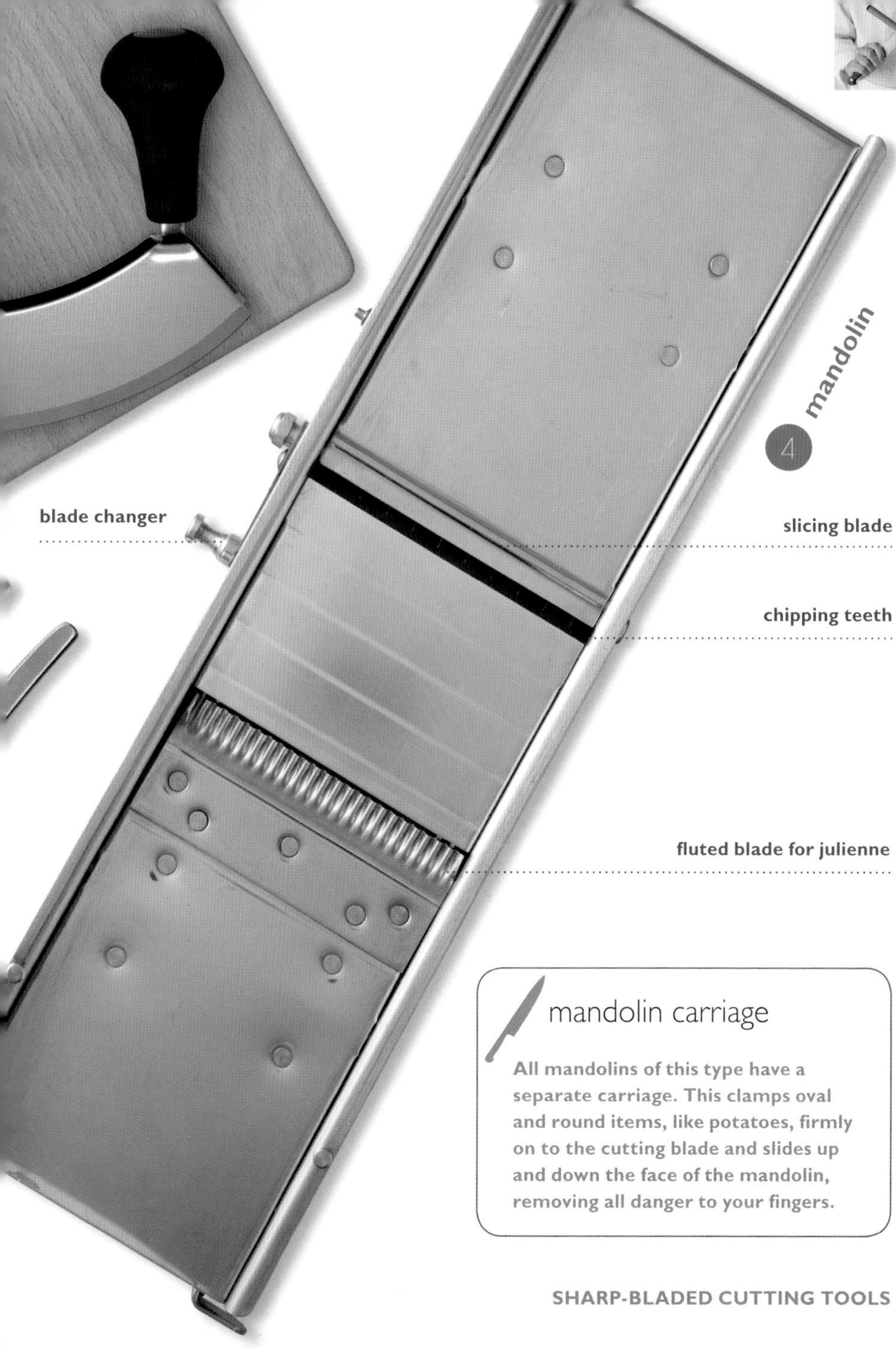

mandolin

4

blade changer

slicing blade

chipping teeth

fluted blade for julienne

mandolin carriage

All mandolins of this type have a separate carriage. This clamps oval and round items, like potatoes, firmly on to the cutting blade and slides up and down the face of the mandolin, removing all danger to your fingers.

these make life much easier in the kitchen, give a **professional** look to your cooking, and are easy to use

1 **set of cutters**

2 **box grater**

3 **nutmeg grater**

4 **microplane grater**

5 **shrimp deveiner**

6 **meat fork**

OTHER CUTTING ACCESSORIES

1 **set of cutters** The smallest cutters punch holes in pastry, eg in tops of pâtés en croûte or steak and kidney pies. Small ones stamp out centres of pineapple slices and large ones are used to shape potatoes into cylinders for lattice potatoes (p71). Medium cutters are useful for mince-pie bases and medium-small ones for cutting out mince-pie tops. 2 **box grater** This traditional grater is always useful; the handle on top enables you to apply pressure to make it stand securely on a plate. The various shredders on the sides are good for grating all types of cheese, citrus zest, and slicing potatoes. Take care of your fingertips and nails. 3 **nutmeg grater** Nutmeg is a volatile spice, so has to be freshly grated. Too much nutmeg gives a bitter flavour, but the sharp teeth on this little grater quickly supply the right amount. 4 **microplane grater** This fine shredder copes with almost every food. The curved blade is made from surgical-grade steel with chemically etched, razor-sharp perforations, creating a grating surface that needs a minimum amount of pressure during use. Models come with coarse or fine holes. A similar, stainless-steel citrus grater (not shown) shreds the skins of citrus fruits finely and evenly, without zest clogging the blade, enabling you to stop when you reach the bitter white pith. Both graters are dishwasher-proof. Take care of fingertips and nails. 5 **shrimp deveiner** The point helps to cut through the flesh on the outside of the shrimp and lift out the dark, intestinal thread. The tool is used widely where shrimp are the size of large prawns. 6 **meat fork** Essential to good carving, the fork pins meat or a bird to the board, while your other hand wields the slicing knife. Also use it to lift and turn roasting birds, as well as hold a roast chicken vertically to tip juices on to a white plate – if they are clear, the chicken is cooked. 7 **Parmesan grater** A simple box, made from plastic or fruitwood, catches and stores the grated cheese, as well as supporting the grater, so you can use plenty of pressure. It is topped with a stainless-steel, easily cleaned blade, with sharp teeth for biting into the Parmesan.

7 **Parmesan grater**

GARNISHING TOOLS

It is easy to collect lots of cutting gadgets. When piled into kitchen drawers, they get in a tangle, can be unhygienic, and hardly ever see the light of day. However, the tools shown here will be used time and time again. Don't put them in the dishwasher.

1 **melon baller** As well as for melons, this is mostly used for other tasks such as coring apples and pears, so choose a good-sized one, with a non-slip handle. 2 **canelle knife** The point in the circular blade carves channels in fruits and vegetables, and takes strips of zest from citrus fruit. 3 **citrus zester** If you tend to shred the ends of your fingers on other graters, this gives a neat and quick way of removing zest from citrus fruits – usually as a first step before chopping. 4 **cheese slicer** Pull the slicer over a hard cheese to take off thin wafers for crostini, open sandwiches, or cheese on toast. 5 **peelers** In a kitchen, nothing is more personal than your favourite peeler. Keep at least one of each type, so that others can help you with the peeling chores. The classic rigid peeler is easier to use than a paring knife for some. A swivel peeler works smoothly at peeling, as well as enhancing the curves of fruits and vegetables, especially pears, apples, and kiwi fruits. The horseshoe peeler is ideal for tough items like squash, celeriac, and swede, and suited to the light touch needed for peeling asparagus. 6 **corer** If used with a powerful Brutus grip (p36) and vertical plunge, this removes apple cores entire. It must be strong, with a non-slip handle. 7 **butter curler** Draw the fluted blade over a slab of butter that is not too cold or hard, and definitely not too soft, to create impressive curls for formal meals. 8 **egg slicers** A wire slicer is perfect for hard-boiled eggs, outdoor sandwiches, and boxed salads. Use scissor slicers for the ultimate party trick: slice off the egg tops, scramble the eggs, and spoon back into the shells with caviar or chives.

4 cheese slicer

3 citrus zester

2 canelle knife

1 melon baller

horseshoe peeler 5

rigid peeler 5

swivel peeler 5

corer 6

butter curler 7

wire egg slicer 8

scissor egg slicer 8

KNIFE SAFETY

A safe knife is a sharp knife and demands the minimum of force if used correctly. The various cutting tasks require different grips so, once you have chosen the knife blade for a task, you need to hold the knife in the most appropriate way. Where possible, cut away from your body. Hold the food with the other hand, so it feeds food into the cutting blade without nicking any fingers.

hand grips

Hold the blade farther away for more dangerous tasks. When cutting delicate items precisely, keep it closer to you. If peeling and trimming small items, hold it closest of all, steadying it with your other hand.

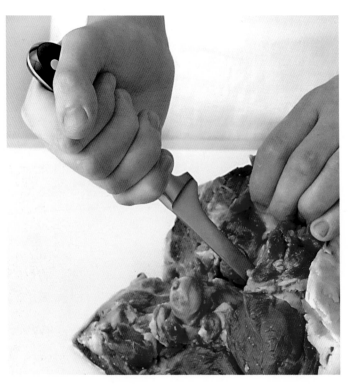

Brutus grip **By holding the knife vertically, as if about to stab, you gain the strength to strip the meat from the bone. The grip also allows you to flex your wrist as you cut and follow the intricate curves of meat bones and joints. Such tasks may also enhance your appreciation of anatomy!**

This powerful grip is the one most used in butchery, where entire carcasses are rendered into joints with little more than a boning knife.

general-purpose grip **Hold the handle firmly in your palm with 4 fingers. The thumb may rest, when necessary, on the spine. The handle must fill your hand for comfort. This grip is good for cleaving tough foods and for general slicing and chopping.**

thumb grip **For precision work, the knife must be an extension of your hand. Squeeze forefinger and thumb on opposite sides of the blade, hold the bolster firmly, and wrap the other fingers around the handle.**

horizontal cutting grip **Press your thumb on the blade, and wrap your fingers round the handle. Hold the blade at a right angle to the work surface and use an even, sawing action, for horizontal and cross-cutting, eg smoked salmon and cakes, and fine slicing.**

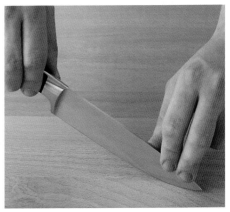

double-handed grip **Ideal for a bunch of herbs: anchor the tip to the board with the fingertips of one hand; hold the handle with your thumb facing you. Chop rapidly while moving the blade in an arc.**

cutting safely

As a general rule, cut all ingredients on a wooden or plastic chopping board, with a damp cloth placed underneath to stop it slipping. Slice an end off round items, so that they sit on the board without rolling.

holding food **To avoid cutting your fingers, grip the food with your hand like a claw, with the fingernails and top finger joints parallel to the blade. You can then safely guide the blade against your knuckles as they move back.**

quick tip

Move any knives resting on the work surface out of your way, but never near the edge of the work surface. If they are lying close to you and the chopping board, turn the blades away from you or, if space is tight, snugly against the board.

cleaning knife blades **Rinse the knife in clean, hot, soapy water. With the blade facing away from you, wipe it with a damp, double-folded cloth to remove any food debris. Wipe from the heel and the spine towards the blade and point – never in the opposite direction. Spritz the blade with sanitizer and dry, using a tea towel and wiping in the same way.**

The most dangerous and common of knife injuries occurs when a person grabs at something in a sink and a blade slices the hand between thumb and forefinger, so never leave knives soaking in soapy water in the sink.

handling knives safely

Treat knives with great respect, especially when others are in the kitchen and when carrying a knife. Never be tempted to try to catch a falling knife – stand out of the way and let it drop to the floor.

pass a knife safely **Hold the knife by the spine and blade, cutting edge downwards. Offer it to the other person at a height that is comfortable for them, ideally over a work surface. Or place on a work top, cutting edge away from both of you, to be picked up.**

tools of the trade

If you do not have a knife holster, and need to transport knives, roll them in a thick tea towel, enclosing the tips and ensuring the blades do not touch each other. Keep the bundle level during the journey. Never leave a cloth over a knife; someone may grab it, unaware of the danger beneath.

walking with a knife **Hold the handle firmly, with the blade close to your side, its tip pointing downwards, and the cutting edge facing behind you. For extra control, place your thumb on the spine. Never run, rush, or otherwise jeopardize the safety of yourself or anyone else. If your kitchen includes a collection of very sharp knives, wear covered shoes to protect your feet.**

STORING KNIVES

Once you have made the outlay on a set of good knives, spare a thought about how to protect the blades. Don't try to store them jumbled in your kitchen drawers or leave them loose. This is unhygienic, dangerous to fingers, and will dull the blades. Opt for a system that allows you to select a knife quickly. Choose a time-honoured material like wood which will hold the blades individually and securely, or a plastic or metal that can be properly cleaned.

storage systems

If space is tight, think of wall storage (see *opposite*). If you have a fully equipped kitchen, you might choose to fit a specially designed knife drawer (*below right*). A knife block is one of the best storage solutions, taking up little space and being completely portable: it can accompany you wherever you work, in the kitchen or al fresco (*below left*).

free-standing knife block **These come in various sizes, so choose one to suit your assortment of knives. Make sure that it is solid and heavy and will not topple over, and that the holes fit your blades. The handles are angled to make it easy to grasp any knife.**

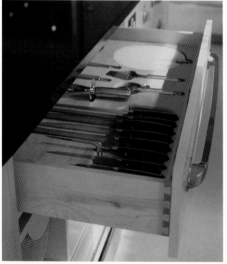

drawer inserts **If you aim for a clutter-free worktop, a sleek kitchen, and safety, lay out your knives in an insert in a drawer. It may be a standard design or custom-made to suit your own collection. This system also avoids blunting the cutting edges.**

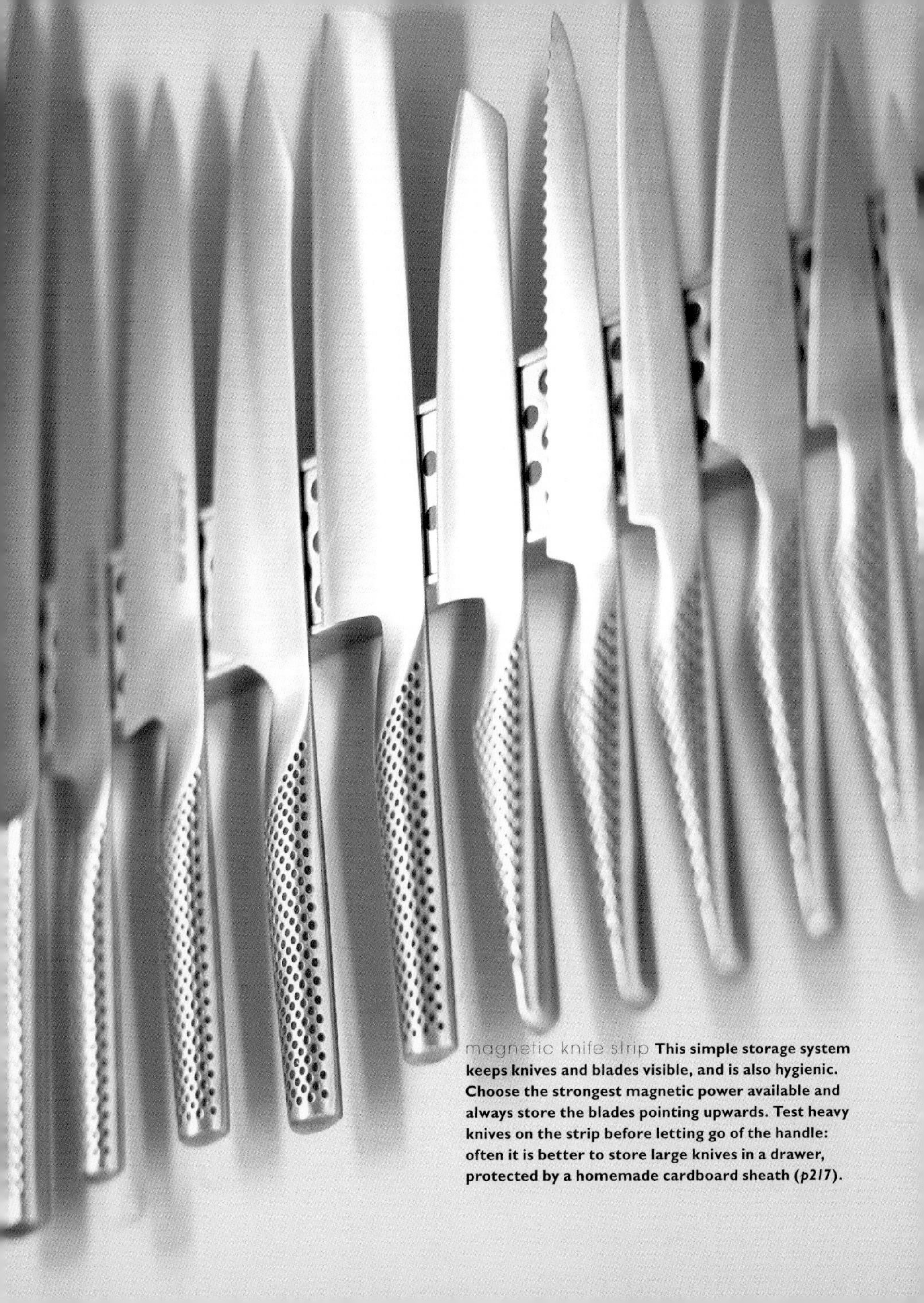

magnetic knife strip **This simple storage system keeps knives and blades visible, and is also hygienic. Choose the strongest magnetic power available and always store the blades pointing upwards. Test heavy knives on the strip before letting go of the handle: often it is better to store large knives in a drawer, protected by a homemade cardboard sheath (p217).**

CARING FOR KNIVES

Keen chefs take pride in sharpening kitchen knives regularly on a stone. There are three main types of finish to a sharpening stone. A rough finish is suitable for completely blunt, chipped, and damaged knives, and a medium finish for sharpening dull knives that have been blunted by daily use. Fine sharpening of knives is done on a superfine finish. If the stone is not completely matt and flat before you begin sharpening, rub it on a fine concrete surface.

general rules of sharpening

Always move the blade across a sharpening stone in one direction, never back and forth or in a circular motion. If using a whetstone, soak it in cold water for 10–15 minutes, until bubbles stop rising. The length of the knife blade defines the angle at which you hold it to sharpen it: blades up to 15cm (6in) long are held at 10°, longer blades at 15–20°.

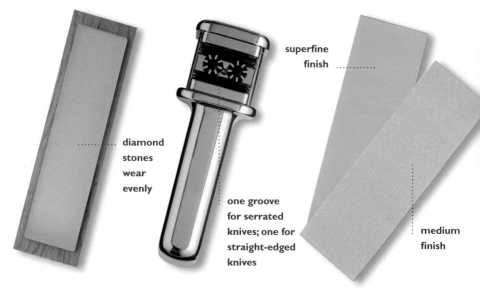

superfine finish ....

.... **diamond stones wear evenly**

one groove for serrated knives; one for straight-edged knives

medium finish

diamond stone **The best stones have 100% diamond crystals and can rapidly sharpen any blade. They do not need to be soaked or lubricated with mineral oil.**

swipe-through sharpener **Blades are drawn through fine grooves, which have sharpening elements at the bases. This is not as precise as a stone.**

combination whetstone **This has 2 or 3 surfaces. Sharpen first on the medium finish, then move to the superfine side. Rubber feet stop the stone slipping.**

sharpening across the stone

Sharpen knives ground on both sides on the right side first, until you feel a burr (rough edge) on the left side. Then sharpen the left side, until you feel a burr on the right. Repeat, 4–5 times, until you have removed the burr from both sides. After sharpening, wash the blade.

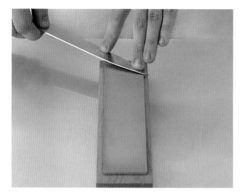

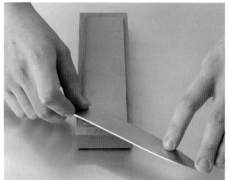

I Place the stone at a right angle to your body. Hold the handle with your right hand, and place the tip of the blade at the end of the stone closest to you. Use your left hand to press the blade firmly and evenly on the stone at the correct angle (see *opposite*).

2 Push the blade slowly over and across the stone, keeping the angle constant. When the blade reaches the other end of the stone, only the heel of the blade should still be in contact with the stone – draw it off the stone in one smooth motion.

3 Turn over the blade and repeat, drawing the blade down the full length of the stone. For a good edge, it is essential to maintain a consistent angle, stroke after stroke, between the blade and the stone. Test if the knife is sharp on a tomato: the blade should slide easily through the skin.

sharpening along the stone

Working in a rhythmic, consistent, and relaxed way keeps the angle and direction of the blade true on the stone – vital for sharpening a blade successfully – and can be done only if you feel balanced. If you find sharpening across a stone (*p43*) uncomfortable, try this method.

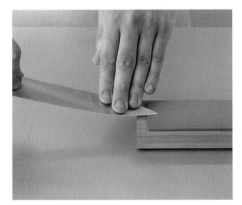

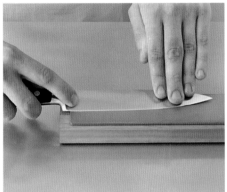

1 Position a sharpening stone parallel to the edge of the work surface and your body. With your fingers, press the blade point on to the top end of the stone, ensuring it aligns with the stone's longitudinal centre.

2 Push the blade along the full length of the stone, while maintaining an even pressure and an angle of 10–20° (*see p42*) between the blade and the stone. Draw the heel of the blade smoothly off the other end.

tools of the trade

Serrated knives that are scalloped only on one side must be sharpened only on the plain side. With knives that are bevelled on one side, make sure that you match the angle of the bevel with the angle at which the edge meets the whetstone.

Alternatively, take your serrated and scalloped knives to a professional sharpening service. Ceramic and Japanese knives are generally best sharpened professionally.

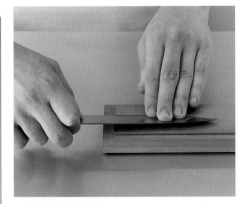

3 Turn the knife over and repeat steps 1 and 2 (*see above*) on the other side of the blade. Take care not to over-sharpen the edge – if the edge is too fragile, it could break and blunt.

honing knives

Steels don't sharpen blunt blades as do sharpening stones, but maintain the cutting edge on a sharpened knife. Keep a steel handy and always hone a knife before you use it. Hone the entire cutting edge from heel to point, otherwise you will grind a curve in the centre of the blade, which will soon stop you chopping effectively.

oval honing steel **This steel is heavier than the round one (see right), but is a treat to use. The oval shape tapers the blade to its original edge and the diamond coating bites into the blade as you stroke it smoothly across. It is not surprising that it delivers a particularly fine result. The average life-span of the steel is shorter, because the diamond coating wears off over time.**

round honing steel **This has a chromium-plated, grooved surface that delivers good results. It is hard-wearing and almost immune to damage. Wipe the surface after sharpening to remove the burr (fragments of blade), which remains on the steel after honing.**

honing on a steady steel

This method is best for beginners. Don't be discouraged if you start by blunting your knives – it happens to everyone. Persevere, work through it, and you will soon be able to maintain a sharp edge on a blade.

1 With one hand, hold the steel (round steel is shown here) at 90° to the work surface, on a cloth to stop it slipping. Take the knife firmly in the other hand, and place the heel of the blade at the top of the steel.

2 Draw the blade steadily down the steel, pulling the knife towards you so that the cutting edge travels across the steel. Maintain the blade at an angle of 10–20° (according to its size, see p42) throughout.

3 Aim to finish the first pass with the tip of the blade at the bottom of the steel, to ensure that all of the cutting edge on one side has been drawn across the steel. Pull the tip smoothly off the base of the steel.

4 Repeat steps 1 to 3 (above and left) with the other side of the blade, keeping a steady pressure and consistent angle between the blade and steel. Repeat the whole process until the cutting edge is honed.

freehand honing

The more you practise this method, the more polished and relaxed you will become. Again, the angle at which you present the knife to the steel is vital – a clanking sound indicates that the angle is incorrect.

I To begin, hold the steel (an oval steel is shown here) firmly in 1 hand. Place the heel of the blade at the top of the steel, with the spine towards your body, at an angle of 10–20° (according to its size, *see p42*).

2 Holding the knife firmly, draw the blade swiftly down and across the steel. Take care to maintain a constant angle between the blade and the steel. Finish the stroke by pulling the tip off the base of the steel.

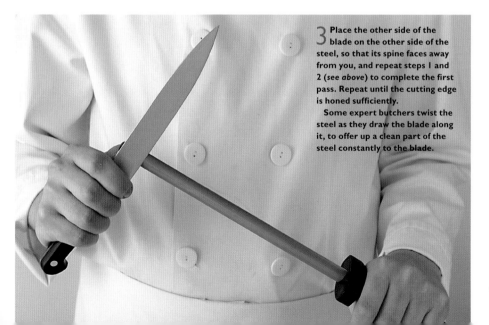

3 Place the other side of the blade on the other side of the steel, so that its spine faces away from you, and repeat steps 1 and 2 (*see above*) to complete the first pass. Repeat until the cutting edge is honed sufficiently.

Some expert butchers twist the steel as they draw the blade along it, to offer up a clean part of the steel constantly to the blade.

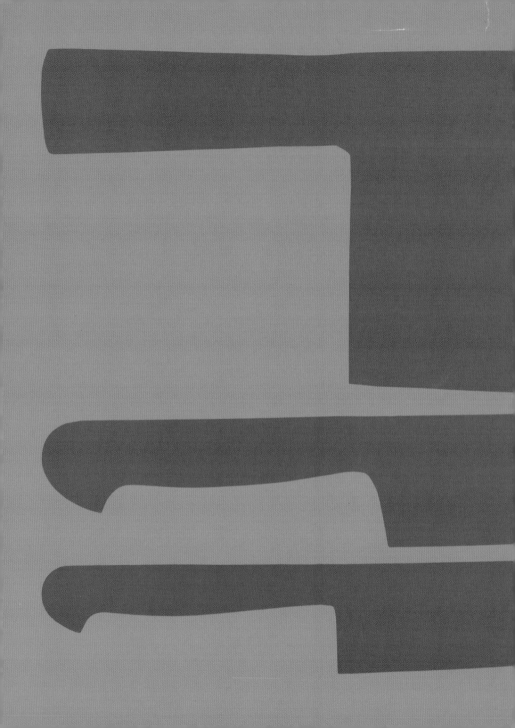

PART 2

KNIFE SKILLS: the application

VEGETABLES

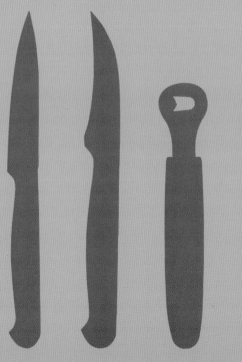

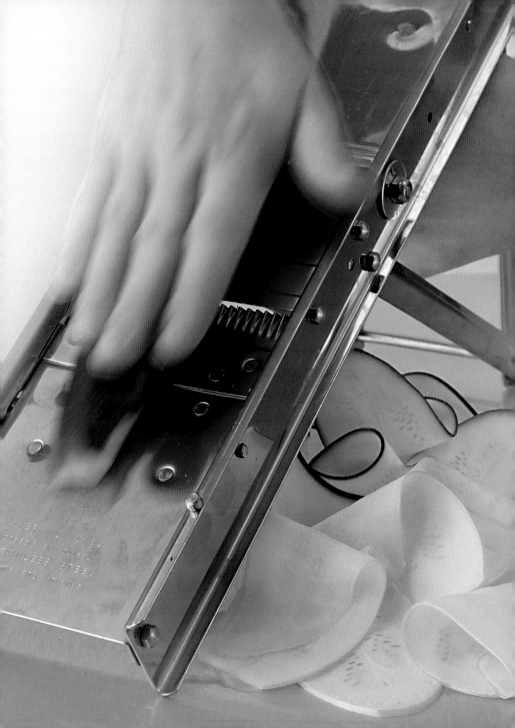

VEGETABLES

Soft and round, hard and oval, tough and fibrous, soft and juicy, or with long stalks and sweetly filled pods, this chapter will lead you through the infinite variety of vegetables, herbs, and fresh spices, and how to deal with them. There are many tools on offer: knives, peelers, graters, canelle knives, and the mighty mandolin, which, with its various cutters, can make you look like a professional chef in a matter of minutes.

preparing vegetables

Knives, on the most elementary level, cut the vegetables into smaller pieces, to help heat reach the cellulose and starches (often the prime component and source of nutrition in most vegetables), making them sweet and soft to eat. Beetroot take well over an hour to cook when whole but, when diced, they can take minutes.

Often vegetables must first be trimmed before cooking, for example cutting flat the stalk ends of cabbages or mushrooms, in order to slice them quickly, safely, and efficiently. And a huge amount of peeling is often necessary.

Use your budding knife skills to give your daily portion of vegetables – so vital for health – an appetizing appearance. Cut leeks and celery into diamonds, slice sugar snaps to expose the baby peas, and shred green cabbage so finely that it glows like jade when blanched. You can also use your cutting skills to produce impressive and colourful garnishes, such as a chilli flower.

Cutting styles of vegetables vary across the cuisines of the world. Slicing vegetables such as spring onions and carrots on the diagonal, with a ceramic or santoku knife, gives an Asian nuance to your food and shortens the cooking time, making them crunchy and bursting with flavour.

The classic French kitchen gives us a legacy of precise shapes and sizes: brunoise, mirepoix, julienne, batons, and batonnets. All are cut exactly to size within a millimetre, but you might prefer a more natural approach, for example shaving parsnips into ribbons for crisps.

cutting for flavour

The fascinating part of vegetable cutting is how we use it to control flavour in our food. If you simmer large chunks of carrot, onion, leek, mushroom, and tomato, they will leach out their flavour slowly but surely, over several days, for a fine stock. If you chop the same vegetables small, they will have greater surface areas and quickly infuse a liquid with flavour. Chopped to a size somewhere in between, fried, coated in flour and roasted, they will colour, sweeten, and thicken a delicious brown sauce.

CUTTING VEGETABLES

To address the disciplines of vegetable chopping, begin with cutting round vegetables, and then tackle tough vegetables like cabbages. Get acquainted with the terminology of the classical kitchen, by learning basic cuts such as mirepoix, batons, batonnets, julienne, and dice. Then you can extend your skills to making vegetable stars and diamonds – food fit for a party.

getting to grips with vegetables

Many vegetables, such as onions, pods, and squashes, are round or awkward shapes and pose a real danger – how to cut them safely with a large, sharp knife, as they roll out of your grasp? The rule for cutting such vegetables is first always to cut a straight side. This is most often achieved by cutting the vegetable (or fruit) in half; place the flat side on the board, and the item is safe to chop or cut.

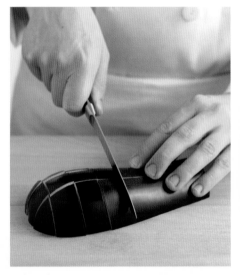

aubergine **Aubergines are soft and don't slip, but their skins are tough and can blunt a tapered ground blade, so use a serrated knife. Cut each aubergine in half; place the cut side on the board; halve lengthways. When slicing further, hold the pieces together, slowly withdrawing your hand before the advancing knife.**

cabbage **Use the largest chef's knife or a cleaver on tough items like this cabbage. Cut the cabbage in half: plunge the knife down the middle, and use your other hand, fingers outstretched, to push the blade through if necessary. Then place each half flat-side down to cut into quarters, applying pressure if needed.**

mirepoix

This is a mixture of chopped celery, carrot, leek, and onion. The size of the vegetables varies from very large chunks to bite size according to the length of time the mirepoix is to be cooked – the longer the cooking time, the larger the vegetable.

large **Use in long-cooking stocks. Cut the celery, carrot, and leek into 5cm (2in) chunks, with a small chef's knife. Cut the onion lengthways into quarters.**

medium **Use in braised dishes and stews. Cut the celery in half crossways, then lengthways. Gather the pieces together and cut crossways into 2cm (¾in) dice. Cut the carrot and leek into 5cm (2in) lengths, quarter lengthways, and cut crossways into thirds. Halve the onion lengthways, then slice crossways into 2cm (¾in) pieces.**

apply this skill

Mirepoix provides a time-honoured way of training young chefs to cut all vegetables to the same size and shape for seamless food preparation. **Aim for a neat appearance and uniformity** if you don't have the inclination or time to do more.

Julienne (*p57*) should be no longer than the width of a soup spoon, to make vegetables in soup easy to sip.

small **Use for garnishing, keeping the cooking time short. Cut the celery, carrot, and leek into sticks. Stack the sticks and cut crossways into 5mm (¼in) dice. Finely dice the onion using the professional method (*p58*).**

cutting batonnets

Batonnets are 5mm (¼in) wide and 5–6cm (2–2½in) long. To prepare them, choose long, straight vegetables (carrots are shown here). Peel each carrot very evenly with a swivel peeler to a smooth, tapering cylinder. Set the aperture of the mandolin blade to 5mm (¼in).

1 Clasp the mandolin tightly with one hand. With the other hand, hold the carrot to the mandolin; grip it carefully with your fingers to slide it up and down for the first slices, until you get some purchase. Then press the carrot on to the blade with your palm.

2 You can now slice the carrot at speed, with no fear of cutting your fingers or the palm of your hand. Stack up the carrot slices, in the order they fell from the mandolin. Square them off: cut them into neat rectangles, all the same size, with a chef's knife.

3 Slice lengthways, using the chef's knife, into parallel strips of equal width, about 5cm (¼in) wide, to make batonnets.

cutting julienne

Julienne are 2.5–5cm (1–2in) long and 3mm (¹/₈in) thick (radish is shown here). Clean the vegetables, eg top and tail radishes, with a paring or turning knife and cut a straight side on each, so it is flat. Julienne may be used for Japanese dishes and garnishes.

quick julienne **The easiest way to make julienne is to use the fluted cutter blade towards the base of the mandolin. Adjust the blade to a fine aperture. Place the radish flat-side down on the blade, and rub it up and down, using the palm of your hand. Take care: this one can bite a little!**

top technique

Another way of making julienne is to set the cutting blade of the mandolin to "very thin", and cut the radish into fine slices. Then stack the slices and, with a small santoku or chef's knife, cut the slices into fine strips.

Alternatively, place the julienne cutter behind the main cutting blade. Rub larger vegetables up and down, for very small batonnets. Take care with your fingers!

dicing

Dicing is a useful solution if you need to eke out a choice ingredient or like to deconstruct a classic recipe (cooking ingredients in different ways). Dice at the last possible moment to preserve the flavour.

1 Wear gloves when cutting beetroot and use a plastic board. Peel and square off each vegetable and cut into a rectangle. Slice into straight batons (larger than batonnet, see p56). Pile the batons on top of each other to make the dicing process as quick as possible.

2 Many other vegetables, eg swede, may be cut into batons; stack, then slice across into equal-sided dice. The sides of the dice should be straight: use a santoku knife for the greatest accuracy.

a sweet vinaigrette is
ideal for salad dressings,
fish, and seafood

apply this skill

Infuse tiny dice, cut from thinly sliced, ripe, aromatic vegetables – celery, red pepper, leek, fennel, and onion – in extra-virgin oil for about an hour to make a delicious sweet vinaigrette. Add a strip of orange zest for a touch of Provençal sunshine.

cutting diamonds

This is ideal for plain, boiled green vegetables, such as leeks (shown here) and runner beans. Cut sugar snaps at a sharper angle, almost tip to tail, into two or three thin slices, to release the sweet, tiny peas.

1 Use a fine slicing knife to cut the leeks lengthways, with the white root ends still attached. Rinse under running water, holding the green leaves downwards. Cut each in half and place the flat side on the board.

2 Cut each half diagonally into diamonds. When you become more practised, you can pile several leeks on top of each other, and slice them with a large chef's knife – it makes the job much quicker.

cutting ribbons

Use the mandolin to create vegetable ribbons, for example of courgettes (p50), of parsnip (shown here), or of daikon for delicious deep-fried crisps. With the peeler of your choice, peel the parsnips lengthways, until smooth.

parsnips Set the aperture of the blade to "fine". For the first ribbon, grip carefully with your fingers. The 2mm-thick ribbon should just hold it shape without falling apart. Press the parsnip on to the blade with your palm, and run it across the cutter blade.

cutting stars

The canelle knife enables you to make attractive stars that give a different look and style to long vegetables, such as courgettes and carrots. Stars are very good for stir-frying.

I With the canelle knife, cut long grooves into the skin of each vegetable (here, courgettes). To get evenly spaced grooves and equally sized scallops, carve the second groove directly opposite the first, the fourth groove opposite the third, and so on.

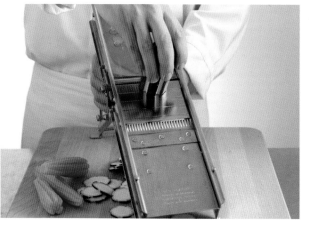

quick tip

Peel carrots evenly until smooth before cutting the grooves. You can also cut the carrot in half lengthways, and slice a piece off the thick end, on the diagonal. Then slice it into thin slices on a mandolin.

2 Set the blade on the mandolin to 3mm ('/8in) thickness. Cut the end of the courgette and place it cut-end down on the mandolin. Hold it firmly and pass repeatedly over the cutting blade to cut the courgette into rounds, or stars. Take care with your fingertips.

ONION FAMILY

Onions are peeled and diced or sliced, then usually sweated. Leeks are usually cooked, whole or cut up, by boiling, braising, or baking in a sauce or vegetable mixture. The pungency of garlic depends on how it is prepared. When left whole, it tastes mild and sweet; when chopped, the flavour is stronger – the more finely it is chopped, the more pungent it becomes.

peeling & dicing onions

top technique

Onions may be sliced into half moons or rings of varying thicknesses.
• To slice half moons, cut the onion in half (see step 1, *left*) and peel it. Lay each half cut-side down on the board and slice across (not lengthways).
• For rings, peel the onion, keeping it whole. Hold it firmly on the board and slice across into rings. Discard the root and stalk ends.

1 Using a chef's knife, cut the onion lengthways in half. Peel off the skin, leaving the root on to keep the onion halves together.

2 Lay one half flat-side down. Make 2 or 3 slices into it horizontally, cutting up to, but not through, the root end.

3 Cut the onion half vertically now, slicing down through the layers, again cutting up to but not through the root end.

4 Cut the onion across the vertical slices to get an even dice. Discard the root end.

washing leeks & cutting into julienne

This mildest member of the onion family often collects soil between its many layers, so needs to be thoroughly washed before use.

1 | With a chef's knife, trim off the root end and some of the dark green leaf top. Cut the leek in half lengthways and fan it open, holding the white end.

2 | Rinse under cold running water to remove the soil from between the layers. Gently shake the leek to remove excess water, then pat dry with kitchen paper.

3 For julienne, cut off all the green part. Cut the white part across into sections of the required length. Lay a section flat-side down and slice into fine strips about 3mm (⅛in) wide.

preparing garlic cloves for roasting

Open the head of garlic by pulling it apart or bashing it with a cleaver or the side of your clenched fist. Whole cloves are ideal for roasting with lamb, beef, or Mediterranean vegetables, or infusing in milk for garlic mashed potatoes.

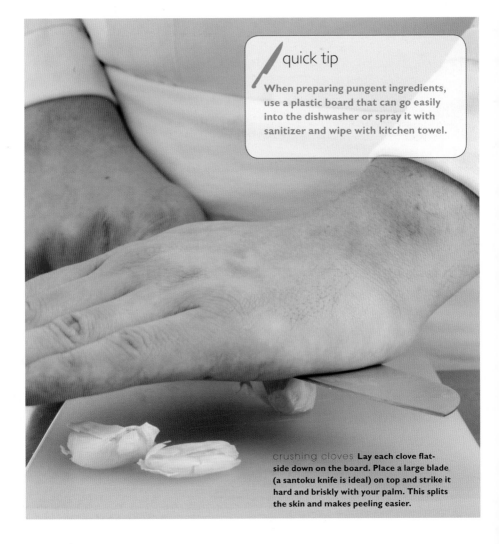

quick tip

When preparing pungent ingredients, use a plastic board that can go easily into the dishwasher or spray it with sanitizer and wipe with kitchen towel.

crushing cloves Lay each clove flat-side down on the board. Place a large blade (a santoku knife is ideal) on top and strike it hard and briskly with your palm. This splits the skin and makes peeling easier.

peeling & chopping garlic

Freshly chopped garlic is an essential part of gremolata – an Italian combination of garlic, lemon zest, and plenty of chopped parsley – and makes a lively topping for grilled sardines. Where you need diffusion of flavour, as in dips, relishes, chutneys, or chilli jam, crush the garlic first.

whole cloves **If you need to keep the cloves whole, use a turning knife carefully to peel the thin, papery skin from the cloves of garlic, one by one. Cloves used whole in cooking yield only a mild flavour of garlic.**

crushing **Line up the board with the edge of the work surface. Cut each clove in half and put flat-side down close to the board edge. Using the spine of a santoku or chef's knife closest to the bolster, chop with small strokes to crush. This gives the strongest flavour.**

chopping **Lay the flat side of the clove against the board. With a santoku knife, cut into 3 (if the garlic is fat enough) lengthways, then cut across into equally tiny pieces. This gives a strong flavour of garlic.**

quick tip

To stop the garlic sticking to the knife as you are chopping it, sprinkle it with a little salt.

To make a garlic paste, chop each garlic clove until it is very fine and press and smash the garlic with the flat of the knife blade on the chopping board.

ROOT VEGETABLES

Vegetables that grow underground include carrots, turnips, potatoes, parsnips, swede, beetroot, salsify, celeriac, radishes, and Jerusalem artichokes. Inexpensive, ubiquitous, and nourishing, they are the ideal vegetables on which to practise your knife skills. Most are peeled, before or after cooking, and their dense textures make them easy to shape and chop.

preparing roots & tubers

In many climates, roots and tubers are familiar vegetables throughout the year. Early in the season when they are tiny, a simple scrape with a paring knife or peeler is all that is required to prepare them for the table, and to keep intact as many of the nutrients as possible. Later, when they have grown into massive specimens, they need cutting and shaping so that they are quick to cook, provide plenty of portions, and are attractive to eat.

In the winter, when vegetables are enormous and their skins are thicker and tougher, it is necessary to resort to more powerful tools. The rigid peeler is traditional, but if you prefer, you could use the stronger horseshoe peeler, which has large rubber handles for a better grip, or a stainless-steel swivel peeler to remove the peel.

With a bit of practice, you can "tourner" large, ungainly roots like carrots, swede, and turnips into elegant, football shapes. Cut giant beetroot into large dice to add vibrant warmth to a winter dish. When par-boiling potatoes for mashing or roasting, it is essential to cut them into equally sized portions so that they cook evenly. You can also cut a variety of chips for frying from beetroot, potatoes, and parsnips.

peeling **A peeler is a very personal cutting tool. Use the one that best suits you and your grip: hold it at an angle that enables you to take the lightest of peelings off the vegetable (salsify is shown here), and to give it a good shape.**

if prepared with skill, **roots** can be **star ingredients** in many dishes

classic French fries

To ensure chips are golden outside, cooked within, and crisp all over, they must be small and all the same size, otherwise some will burn and taste bitter. Always choose a potato that is good for frying – the larger the better. After peeling, keep them in water until you are ready to cut them into chips, but ensure they are perfectly dry before deep-frying. For hand-cut, classic French fries, aim to cut them into 5x5mm (¼x¼in) square batonnets (p56) that are about 6cm (2½in) long.

quick French fries **The quickest way to make classic French fries is to use the mandolin. Adjust it to bring the large julienne cutter up behind the main blade. Square off each potato, ie trim it until all the sides are rectangular, and place it in the carriage. While pressing the potato firmly downwards, pass it up and down the mandolin and over the blade.**

quick tip

If you don't have a mandolin, begin with Pont Neuf fries. Cut 7.5cm (3in) long and just over 1cm (¹/₃in) square, these are the largest chips of all. Some chefs boil them first, while some fry them three times, so that they are dry and floury within, but still crunchy.

pommes allumettes & pommes pailles

Pommes allumettes are much thicker than pailles. They have square-cut ends, like matchsticks, whereas pailles have naturally tapered ends, like straw. If you don't have a mandolin, you can cut them with a knife, but make sure that you cut to the same dimensions for even cooking.

pomme allumettes **Square up each potato and place in the mandolin carriage. Turn the small wheel on the side to fit the julienne cutter under the main blade. Push the potato firmly up and down to pass over the blade and create "matchsticks", that are about 3x3mm (⅛x⅛in) square and 4cm (1½in) long.**

pomme pailles **Do not trim or square off the peeled potato, but use a mandolin to cut it into very thin slices (p70). Pile the potato slices on top of each other. Then, with a chef's knife, cut them into very thin strips, less than 2mm wide, so that you produce straw-like chips.**

finished fries **You might have to fry pommes allumettes twice to get the matchsticks stiff and crisp. Use groundnut or vegetable oil at a temperature of 160–180°C (320–356°F). Drain on a kitchen towel and serve with roasts and grills.**
   Fry pommes pailles only once, a small spoonful at a time. Drain on a kitchen towel. They will curl, to create wonderfully crisp and fragile fries that are excellent served with seafood.

slicing thinly

A mandolin will deliver precise and uniform slices of potatoes and other roots and tubers. It can also be used to slice other dense vegetables, such as winter squashes.

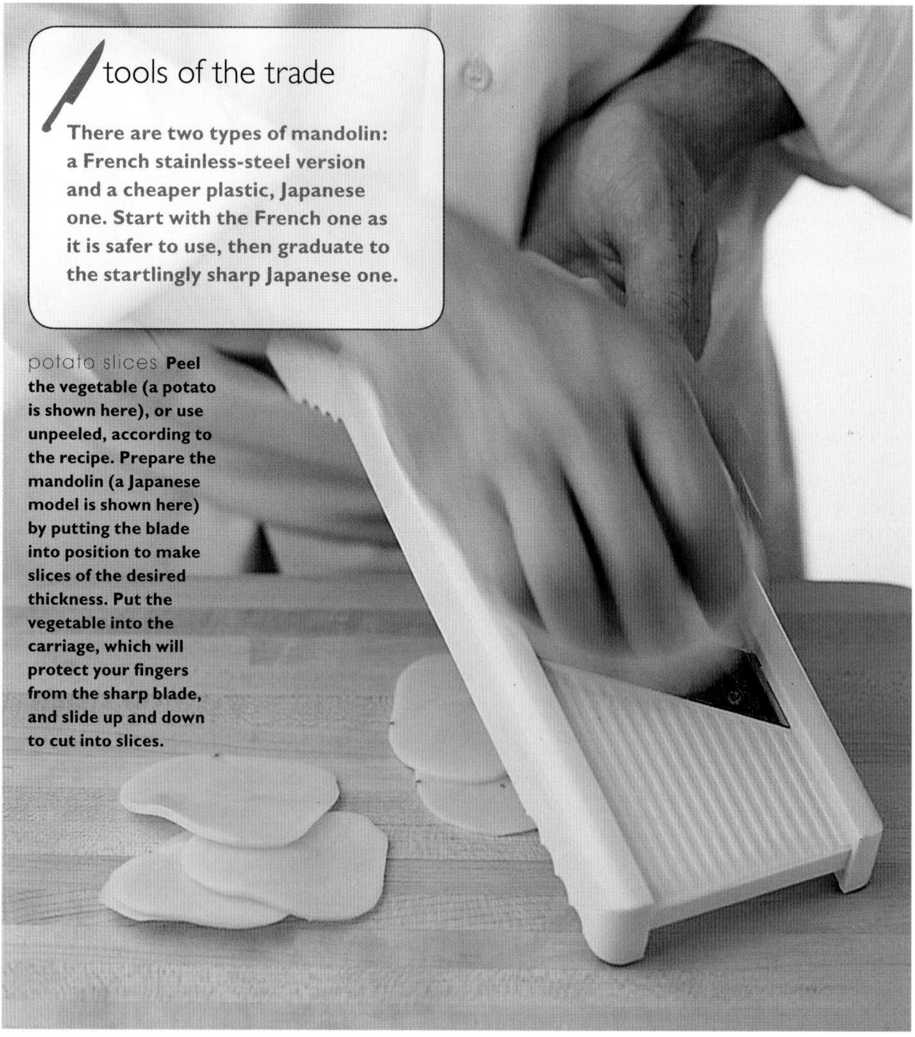

tools of the trade

There are two types of mandolin: a French stainless-steel version and a cheaper plastic, Japanese one. Start with the French one as it is safer to use, then graduate to the startlingly sharp Japanese one.

potato slices **Peel the vegetable (a potato is shown here), or use unpeeled, according to the recipe. Prepare the mandolin (a Japanese model is shown here) by putting the blade into position to make slices of the desired thickness. Put the vegetable into the carriage, which will protect your fingers from the sharp blade, and slide up and down to cut into slices.**

lattice potatoes

For lattice potatoes, also known as pommes gaufrettes, peel large potatoes and keep in cold water until needed. Cut the top and bottom off each potato, then cut into sections a little bigger than the cutter.

1 Stamp into each potato section with a large, round cutter 5.5cm (2⅛in) diameter; it will get wedged in the potato. Cut away the excess with a utility knife, using the cutter as a guide to produce a perfect tube.

2 Push a flat end of the potato tube over the julienne teeth of the mandolin. Discard this first slice. Turn the potato at a 90° angle, and push over the teeth again. This next slice will have lattice-work holes.

3 Repeat step 2, pushing and twisting the potato, until it is all cut into lattice-work slices. Keep in cold water until required. When ready to fry, dry thoroughly and carefully. Deep fry in small batches in groundnut or vegetable oil at 160–180°C (320–356°F), until golden. Lift with a skimmer and drain on a kitchen towel. Serve with drinks or with roast game.

cutting into batonnets & julienne

This method is suitable for cutting round roots and tubers, such as beetroots (shown here), turnips, swede, and celeriac, into julienne strips. You can use the same technique to cut batonnets.

1 Wear plastic or rubber gloves when preparing beetroot to minimize stains on your hands. Peel the vegetable thinly using a vegetable peeler or small paring knife.

2 With a chef's knife, trim the sides of the vegetable to make a square shape.

3 Holding the block gently but firmly, cut into equal slices – 3mm (⅛in) thick for julienne and 5mm (¼in) thick for batonnets.

4 Stack the vegetable slices, a few at a time, and cut into neat sticks that are the same thickness as the slices. Use bleach to clean the chopping board.

turning

Called "tourner" in French, this is a common preparation technique for vegetables in the root and tuber family, such as carrots (shown here), turnips, and potatoes, as well as for summer squashes. The vegetable is shaved into classic, seven-sided football shapes.

1 Peel the vegetables, if necessary, then use a chef's knife to cut them into pieces that are 5cm (2in) long.

2 Holding one vegetable piece between thumb and forefinger, start shaving off the sides of the piece with a turning knife to curve them.

3 Continue cutting while turning the vegetable piece in your hand, to create a football shape with 7 curved sides.

top technique

Turning vegetables takes years of practice, but the result draws gasps of admiration at dinner parties. The skill is in the rhythm: check your weight is evenly distributed over your feet and that your hands are relaxed. Turn the vegetable gently into the hand that wields the knife. Keep practising and if at first you don't succeed, try again.

LEAFY VEGETABLES

Vegetables from this family are often used in salads. The most common of these is lettuce, but there is also a wonderful variety of other salad leaves in all shapes, sizes, and colours.

To prepare salad leaves, trim off the ends of the leaves, discarding any that are discoloured. If the midribs are tough – typically, in large cos, frisée, and radicchio – cut them out (see *opposite*).

chiffonade

The fine and lacy, leafy filaments created by this technique quickly soften in the heat of a pan, but still give a good texture to a soup or vegetable dish. As well as lettuce, herbs with plain, large leaves lend themselves to this method of cutting. Chiffonade of gem lettuce leaves may be used to garnish lettuce soup.

1 **Lay the lettuce leaves on the chopping board, one on top of the other. Roll them towards you – like a cigarette – being careful not to bruise the leaves.**

2 **Using a large chef's knife, cut off the frilly ends of the leaves and discard. Slice the leaves into fine strips using the horizontal cutting grip (p37), with the knife blade positioned safely against your knuckles.**

quick tip

A generous heap of sorrel chiffonade is essential to sorrel sauce, a cherished accompaniment to salmon. Sprinkle a chiffonade of basil leaves over salads, soups, and pasta, for additional flavour and colour.

trimming & slicing hearty greens

Before cutting a chiffonade of most hearty greens, such as Swiss chard, kale, spring greens, and the leaves of turnips and beetroot, they are trimmed to remove the central rib. Chinese crispy "seaweed" is often a deep-fried, fine chiffonade of cabbage or spring greens.

Discard all limp or discoloured leaves. Using a chef's knife, quickly slash each leaf on either side of the rib. Remove and discard it. Rinse the leaves in water and pat dry in a tea towel or with kitchen paper.

2 Grab a handful of leaves and roll loosely into a bunch. Cut across the roll into strips of the desired width.

top technique

With more tender leaves, such as spinach and sorrel, you can just pull off the stalk and rib instead of cutting it out with a knife. Spinach tends to trap soil so, if not pre-washed, immerse the leaves in several changes of cold water to clean it. After draining in a colander, pat dry in a tea towel or with kitchen paper. Fold each leaf in half and pull off the central rib and stalk. The leaves are now ready to eat raw or to cook.

CABBAGE FAMILY

The most common way to prepare tight, round heads of cabbage is to quarter them and cut out the hard central core, then to shred coarsely or finely. Loose-leaf heads, such as Chinese cabbages and leaves, are prepared more like hearty greens (p75). The usual method of preparing broccoli and cauliflower is to trim the stalk and separate the florets, but heads are sometimes cooked whole.

coring & shredding cabbage

1 Holding the head of cabbage firmly on the board, use a large chef's knife to cut it lengthways in half, cutting straight through the stalk end.

2 Cut each cabbage half lengthways in half again, with a small chef's knife, cutting through the stalk end. Cut out the hard core from each quarter.

3 Take one cabbage quarter and place it cut-side down on the board. With the large knife, cut across the cabbage, creating shreds of the desired thickness.

top technique

Cabbage can be quickly shredded by hand with a chef's knife (see left). For the most efficient action and control, keep the point of the knife on the board as you raise and lower the handle; guide the knife blade with the knuckles of your other hand. A shredding disc on a food processor is faster, but the shreds will not be as pretty as hand-cut ones.

preparing broccoli florets

1 Lay the head of broccoli flat on the chopping board. With a chef's knife, cut off the thick portion of the stalk, cutting just below the floret stalks.

2 Remove the florets by sliding the knife between their stalks to separate them. Rinse the florets in cold water and drain in a colander.

preparing cauliflower florets

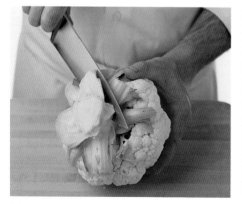

1 Lay the head of cauliflower on its side on the chopping board. With a chef's knife, cut off the end of the stalk. Pull or cut off any leaves.

2 Turn the head core-side up and use a paring knife carefully to cut the florets from the central stalk. Rinse the florets in cold water and drain in a colander.

SQUASH FAMILY

Hard-skinned winter squashes, such as pumpkin, butternut, and spaghetti, need more preparation than summer squashes. Soft-skinned summer squashes, like courgettes, cucumbers, vegetable marrow, and pattypan, may be eaten unpeeled, and may be cooked whole or cut attractively.

halving, seeding & peeling winter squashes

The skin of winter squashes is thick and woody, and the seeds are fully developed. Because of this, winter squashes are always peeled (either before or after cooking) and the seeds and central fibres are removed. Use a sharp, heavy knife for cutting. Butternut squash is shown here.

1 Hold the squash firmly; with a chef's knife, cut into half from the stalk end directly through the core.

2 Use a spoon or a small ice-cream scoop to remove the seeds and fibres from each squash half. Discard the seeds and fibres.

3 Cut the squash into sections. If removing the skin before cooking, peel the sections using a vegetable peeler or small paring knife.

cutting batonnets

This method of cutting batonnets is also suitable for other long, slim vegetables, such as carrots, parsnips, and salsify. A courgette is shown here.

1 Cut off both ends of the courgette, then cut it in half lengthways. Cut each half again to make slices 5mm (¼in) thick.

2 Lay each slice of courgette flat on the board and cut across it to make sticks or batonnets about 5mm (¼in) wide.

quick tip

Slicing vegetables evenly and thinly lengthways, as shown here, takes plenty of practice and can be time-consuming. Courgettes are ideal for practising on, but if you don't have the time, use a mandolin (*p31*) instead, which will give you long, even slices.

making cucumber cups

scooping **Cut one of the cucumbers into about eight 2.5cm (1in) pieces. Using a round cutter or a paring knife, remove the skin. Hollow out the inside of each cucumber piece using a spoon or melon baller to make a "cup". Fill with a quick salmon mousse or, Asian-style, with little prawns in a sesame, garlic, and chilli marinade.**

PODS & SEEDS

Unripe pods, such as green beans, mangetouts, and sugarsnap peas, need little preparation beyond topping and tailing (see *opposite*). The tough strings along one or both sides of mangetouts and some green beans, such as runner beans, need to be stripped off before

cooking, and this can be done by hand (*p82*). Large beans and mangetouts can be cut into decorative shapes, such as diamonds (pp82–3). Although not in the same botanical family as peas and beans, sweetcorn is treated in a similar way in the kitchen after preparation.

cutting off sweetcorn kernels

After removing the husks and silk, the whole ear of corn may be boiled, and the kernels served on the cob, or the kernels can be cut off to be boiled, steamed, sautéed, or simmered or baked in a sauce or soup.

I Pull off the husks and all of the silk from the ear of the corn on the cob and discard.

2 Hold the ear upright on a cutting board and, using a chef's knife, slice straight down the sides to cut the kernels off the cob.

topping & tailing beans

Helping to top and tail peas and beans on summer afternoons is a fond childhood memory for many, but modern living often requires faster action, so use kitchen scissors to trim a bunch at a time.

trimming **Grasp a bunch loosely. Shake upside down on the work surface until all the heads line up perfectly. Snip off the tips with the scissors, taking care to remove just the tips so the beans retain their elegant shapes and don't look stubby. Turn the bunch the other way round and snip off the tails.**

preparing peas & beans

Fresh beans and peas should break with a brittle snap, making them easier to trim and cut. Check for strings before cutting or cooking. Some long runner beans have no strings, but smaller beans and mangetouts do. Cut large beans and mangetouts into diamonds (*p60*) with single, sharp, oblique cuts, using a santoku or a similar knife.

removing strings
Carefully tear off the tip, which will be attached to the string, then pull it along the side to remove the string. If you are preparing beans that have a tough string on the other side too, carefully tear off the tail of the bean in that direction and pull off the attached string.

preparing vegetables for stir-fry

Peas, beans, and corn are great stir-fry ingredients, adding sweetness and bright colour, alongside other vegetables. For the most attractive presentation, be prepared to devote some time to this task. Trim all the vegetables carefully. Then cut, chop, and slice them finely to give the largest surface area, so that they cook quickly but remain al dente. It also means they can absorb the flavour of the oils, aromatic ingredients, and spices. For example, cut beans into diamonds but leave bean sprouts whole, use the canelle knife on carrots (*p61*), and cut spring onions lengthways before slicing them diagonally.

tools of the trade

This is a chance to appreciate your best knives. Bring out the Japanese vegetable, ceramic, and santoku knives. The knives should be honed before you begin so the cutting edges sink effortlessly through the softer vegetables. Enjoy the crunch as you chop mangetouts.

french beans and
mung bean sprouts
look **great** as part
of a rainbow stir-fry

top technique

There is an order to wok
cooking: the aromatics
(eg spring onion) go in
first to give the textures
and fragrance unique to
Chinese cooking. Speed
and control are vital,
so line up the vegetables
in sequence to obtain
the best results.

FRUIT VEGETABLES

Tomatoes, aubergines, peppers, chillies, tomatillos, and avocados, which are fruits according to botanists, are treated as vegetables in the kitchen. All are straightforward to prepare, but take care when stoning avocados – never remove the stone from the knife with your fingers.

halving & stoning avocado

1 **Using a chef's knife, slice into the avocado, cutting all the way around the stone.**

2 **Twist the cut halves gently in opposite directions to separate them.**

3 **Strike the stone with the knife blade to pierce it firmly. Lift up the knife to remove the stone from the avocado half.**

 quick tip

The flesh of avocado discolours quickly when exposed to the air, so serve promptly or rub or toss it with lemon or lime juice.

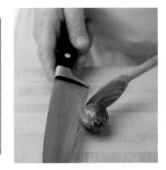

4 **Use a wooden spoon to carefully pry the stone off the knife. Discard the stone.**

5 **Holding an avocado half in your hand, gently scoop out the flesh with the help of a rubber spatula. Repeat with the other half.**

peeling & dicing avocado

| After cutting
the avocado in
half and removing
the stone (*see
opposite*), use a
paring knife to
cut the half into
quarters and
remove the peel.

2 **Using a chef's knife,**
cut the avocado
flesh into neat slices.
To dice the flesh, slice
thinly and then cut
across the slices to
make dice.

preparing sweet peppers

Cut green, orange, yellow, and red peppers into squares for roasting, julienne for stir-frying, dice for brilliant colour and sweetness with fish and salads, or slice off the tops and core and fill them for baking.

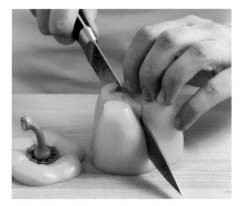

1 Place the pepper on a board and cut off the top and bottom with a utility knife. Stand the pepper on one end, hold it firmly, and cut in half lengthways. Winkle out the core and seeds with the point of the knife.

2 Slice the pepper into manageable portions. To cut out the ribs, lay a section of pepper flat. Hold the knife, with thumb on bolster (p37), horizontal to the chopping board and cut off the pale, fleshy ribs.

3 Cut into smaller sections, following the natural divisions in the pepper, then chop into batonnets (p56) or julienne (p57), depending on the size of the pepper and the dish that you are preparing.

decoring for stuffing and roasting **Cut around the stalk of the pepper and pull it off, taking the core with it. Rinse inside to remove all the seeds, then blot dry with kitchen paper.**

chilli peppers

If preparing a lot of chillies, wear plastic gloves, or they will burn you for hours afterwards. If you don't, wash your hands and avoid touching any tender part of your body, including eyes and lips, for several hours.

chopping chillies
Using a paring knife, cut off the top and bottom of the chilli. Cut in half lengthways. Winkle out the core and seeds with the tip and point of a knife. As the skin is tough, place the chilli fleshy-side up on the board, and slice lengthways into fine strips. If required, the strips may then be diced (p58).

making chilli-flower garnishes

I With the paring knife point, cut through from the stalk to the tip. Rotate the chilli 90°, and cut again. Flatten and tease out the long quarters; cut each in half again. The chilli is now divided into eight "petals".

2 Remove the ribs to help the petals curl; don't worry if some seeds remain. Place in a bowl of iced water for several hours, and allow to curl. Use for garnishing canapé plates, rice, and Asian-style dishes.

peeling, seeding & chopping tomatoes

When tomatoes are used in a soup or sauce that is not passed through a strainer, they are often peeled and seeded first. This peeling method is also used for fruit, such as peaches and plums, and for chestnuts.

1 With the tip of a paring knife, cut around the core of each tomato to remove it. Discard.

2 Score an "**X**" in the skin on the base of the tomato. Immerse it in a pan of boiling water. Leave in the boiling water for about 20 seconds, until the skin splits.

3 Lift the tomato out of the pan of boiling water and then immediately submerge it in a bowl of iced water to cool.

4 Pull off the skin with your hands and the help of the paring knife. Cut the tomato in half and gently squeeze out the seeds.

5 Place each half of tomato cut-side down on the board and cut into strips. Then cut across the strips to make dice (known as tomato concassé).

chopping button mushrooms

Button mushrooms give you an opportunity to improve your cutting skills, as they don't slip on the board. Diced mushrooms, called duxelle, add sweetness, eg to a white wine sauce or beurre blanc.

I Wipe the mushrooms with a cloth – don't wash them. Slice off the stalks with a utility knife; this longer, thinner knife gives greater accuracy than a small paring knife.

With the flat, trimmed sides to the board, cut the mushrooms into fine, 2mm slices.

If preparing dice, stack the slices on top of each other, and cut into thin slices lengthways.

2 Slice across into tiny dice. Use the pinch grip to hold the knife and use a rocking motion, with the blade close to the board, for economy of movement. Hold the mushroom in place with two fingers on top.

quick tip

Dried fungi can add intense flavours to a dish. Mushrooms such as cèpes and shitakes are often used in French, Chinese, and Italian cooking. Soak the dried mushrooms first in warm water, to allow any sand to sink to the bottom. After soaking them, squeeze out all the water, then slice the mushrooms finely, as in step 1.

SHOOTS & STALKS FAMILY

This family includes globe artichokes, asparagus, Florence or bulb fennel, cardoons, and celery. With the exception of globe artichokes, most require little preparation apart from stringing to remove the long, stringy fibres from the outer stalk or bulb (p93).

trimming globe artichokes to serve whole

The hairy "choke" in the centre of a globe artichoke can be scooped out after cooking or before eating. When cut and exposed to the air, artichokes will quickly discolour. To prevent them from browning prior to being cooked, drop them into a bowl of water acidulated with lemon juice or rub all the cut surfaces with lemon.

1 Holding the stalk, cut the tough tips from the artichoke leaves with sturdy kitchen scissors.

2 Use a chef's knife to cut off the stalk flush with the base so that the artichoke will sit upright.

3 Cut off the pointed top. The artichoke is now ready to be cooked.

preparing artichoke bottoms

When all the leaves are removed, what is left is the fleshy, cup-shaped bottom or heart, which is completely edible apart from the central hairy "choke". The bottom can be cut up for cooking or left whole. If served whole, it will be easier to remove the choke after cooking.

1 Pull away or cut off all of the large leaves from the artichoke. Then cut off the stalk flush with the base, using a chef's knife or a long serrated knife.

2 Cut off the remaining soft "cone" of leaves in the middle, cutting across just above the choke (you will see the hairy fibres).

3 Keep rubbing the exposed flesh with lemon juice to prevent browning. Neaten the bottom with a paring knife, trimming off all the remaining leaves and trimming the base so it is slightly flattened.

4 If cutting up the artichoke for cooking, scoop out the choke with a small spoon. Take care to remove all of the hairy fibres. Rub the exposed surface of the hollow generously with lemon juice.

take the **thinnest of peelings**
from asparagus spears and
cook **lightly** to preserve
their delicate flavour

preparing asparagus & celery

Asparagus needs little preparation, apart from trimming off the stalk ends, but you may want to peel larger spears. Celery needs stringing before being cut into shapes, such as batonnets (*p56*) or diamonds (*p60*).

I With a chef's knife, cut the hard ends from the asparagus spears.

2 Holding the tip of a spear gently, use a vegetable peeler (rigid peeler shown here) to peel off a thin layer of skin from the stalk, rotating to peel all sides.

stringing celery **The thick, outer stalks of celery have long, stringy fibres that are best removed. Separate the stalks from the bunch and wash. Then, using a vegetable peeler, peel off a thin layer from each stalk to remove the strings. Florence or bulb fennel and cardoons also have strings that are usually removed in the same way.**

quick tip

Such is the tough, fibrous nature of asparagus peelings that it is wise to avoid throwing them down the waste disposal. The peelings can entwine irreparably around the grinding mechanisms, and bring the entire mechanism to a dead stop.

HERBS

Cooking with herbs is rewarding as they give generously of their fragrance as you chop and slice. Try to prepare them at the last moment, just before you use them, to maximize their fresh flavours. Also, at certain times of the year, a few herbs can go black once chopped.

slicing herbs

Some herbs are sliced and others are chopped. The herbs with soft, large, or long leaves, such as chives, marjoram, basil, sorrel, spinach, and mint are sliced, while rosemary, thyme, coriander, and parsley may be chopped.

I Pick the leaves (basil is shown here) off the stems and pile each leaf on top of the other neatly and gently, without bruising the leaves. The larger the pile, the quicker the preparation will be.

2 Use a chef's knife and a rocking motion, holding the top of the blade with your thumb and forefinger, to slice through each pile. There will be no resistance, so focus on slicing as finely as possible.

rough chopping

This is the best way to tackle big bunches of curly or flat-leaf parsley and coriander. Use a large knife or mezzaluna (see opposite).

fine herbs Cut off the stems and chop the leaves roughly. Hold a large chef's knife at both ends and chop with a rapid up-and-down action, brushing the herb repeatedly into a heap with the knife.

using a mezzaluna

A mezzaluna, whether single- or double-bladed, is a good tool for preparing large bunches of chervil, mint, and coriander and quantities of peeled garlic cloves. This is a tireless method of chopping.

apply this skill

Use the mezzaluna to prepare herbs in garnishes and dressings. For example, chop plenty of flat-leaf parsley with anchovies, black olives, capers, lemon zest, and red onion to make a great topping for roast fillet of cod or croûtes for a drinks party.

chopping with ease **Place the herbs on the board. Keeping your hands relaxed, rock the mezzaluna backwards and forwards, until the herbs are chopped to your liking.**

SPICES

As the flavour cache of the kitchen, spices have the power to transform a meal. The huge variety – both tender and tough – of pods, stalks, barks, stems, roots, and rhizomes can be a challenge for any cook, but fortunately there is a cutting tool for each and every one.

bruising spices

For a steady, gentle release of flavour over a long cooking period, it is best to part-crush tough and stalky spices, using a technique known as bruising. Chopped spices can be too potent, while keeping them whole makes them too mild and may not add enough flavour.

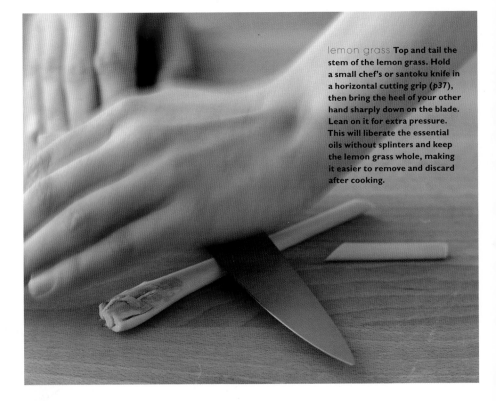

lemon grass **Top and tail the stem of the lemon grass. Hold a small chef's or santoku knife in a horizontal cutting grip (p37), then bring the heel of your other hand sharply down on the blade. Lean on it for extra pressure. This will liberate the essential oils without splinters and keep the lemon grass whole, making it easier to remove and discard after cooking.**

grating spices

Whether dry or juicy, just a touch of a spice can enhance a recipe. Grating spices, preferably using a stainless-steel microplane grater, makes it easier to prepare the correct amount.

ginger **Peel a "thumb" of ginger with a potato peeler or paring knife, then rub it vertically on the razor-sharp grater to render a thick pulp.**

nutmeg **Freshly grated is the only way to do justice to this evocative spice. Grate only what you need. Pick up the fine dust with the tip of a paring knife.**

> ## quick tip
>
> The santoku knife, with its flat blade, rounded tip, and down-turned spine, is ideal for scooping up fine pieces of spice, as well as for the many chopping and slicing requirements of Asian cooking.

extracting vanilla seeds

When vanilla pods are oily and pliable, it is easy to scoop out their seeds. If they are older and dry, massage them with your palm. Empty pods can be rinsed, air-dried, and stored in a cool dark place, and then used again.

1 Cut the vanilla pod in half lengthways with the tip of a turning or paring knife. Slice cleanly and quickly to keep the seeds in place.

2 Using a teaspoon or point of a paring knife, scrape off the black seeds, keeping the paste well clear of your fingers. Drop straight into milk or cream.

FISH & SHELLFISH

FISH & SHELLFISH

Fish and crustaceans are different from mammals and birds because the medium in which they move is denser, but the effect of gravity is much reduced. They need to push against and through water, using a power pack of white muscle. The supporting density of the water enables them to carry these huge muscles without an elaborate and weighty skeleton. This makes killing and preparing fish and crustaceans extremely simple.

Some sea creatures wear their skeletons around them, like lobster, crab, and prawns. If you break through this suit of armour, the white meat within is sweet, boneless, and juicy. Many crustaceans shed their outer shells as they grow, hiding themselves until new, larger shells harden around them. Soft-shell crabs are taken at this stage and prepared alive as a great delicacy.

Bivalves such as oysters, clams, and mussels don't have any skeletons at all, but use a single muscle to open or close their shells for feeding. The shell is able to withstand tons of sea water, crashing and heaving constantly over them.

raw fish & seafood

With the correct cutting tools, you can clean fish to cook whole or strip it off the bone to make fillets. Round fish can be gutted and boned, using knives that are light, sharp, and flexible, and then stuffed with a filling. Long fillets of flat fish can be skinned and divided, rolled, and tied into knots, to make smaller, intriguing portions.

Raw seafood is a great delicacy. Clam and oyster knives, pliers, and tweezers are used in combination with knives to scrape or winkle the creatures out of their shells and skins. Seafood becomes very toxic after killing, so they are killed and prepared just before or during cooking and there is a wide range of cutting tools to crack open shells, dispatch, trim, pierce, and bone.

In Japan, fish is considered only truly fresh when it is eaten raw. The famous sashimi and sushi are prepared with long, pointed, fine Japanese knives such as the *tako hiki* or *yanagi ba*. Raw tuna, squid, and salmon are becoming popular outside Japan but, if you cannot obtain very fresh fish locally, you could cut paper-thin slices from frozen fillets.

cooked fish

The texture of fish changes surprisingly during the cooking process. The flesh has short muscle bundles and very little connective tissue, so it simply falls apart. All you need to fillet a cooked fish is a table knife and spoon. It is a good idea to slice a large fish before cooking, for a neat and appetizing presentation, especially for parties.

ROUND FISH

Round fish are "fin fish" that are round in body shape and have eyes on both sides of their heads. The preparation techniques vary depending on how you are going to cook it. The most common are gutting, scaling, boning (if you intend to stuff the fish), cutting into steaks, and filleting.

gutting through the stomach

Gutting a fish means to remove all the viscera (everything in the stomach cavity). The most common way of gutting fish is to remove the viscera through a cut into the stomach, but fish can also be gutted through the gills (see *opposite*). A pike is shown here.

1 **Hold the fish firmly on its side** and, using a fish knife, small chef's knife, or kitchen scissors, make a shallow slit in the underside, cutting from the tail end to the head end.

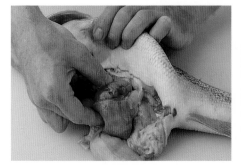

2 **Pull out the guts (viscera), then cut off the gills (see** *opposite*)**, taking care as they can be sharp. Discard the guts and gills.**

3 **Rinse the cavity under cold water to remove any remaining blood and guts. Pat the fish dry with kitchen paper. It can now be scaled (p104) and boned.**

gutting through the gills

This technique is often used for fish to be poached whole or cut into steaks, as well as for small flat fish because it keeps their natural shape. Before gutting this way, scale the fish and trim the fins (*p104*). A rainbow trout is shown here.

1 First, cut off the gills at the base of the head with kitchen scissors. (The gills are sharp, so hook your index finger around them to pull them out.)

2 Put your fingers into the hole left by the gills and pull out the viscera.

3 Using the scissors, snip a small slit in the stomach at the ventral (anal) opening near the tail. Insert your fingers and pull out any remaining viscera.

tools of the trade

Many of us learn to use scissors before we learn to use knives. They are useful when cleaning fish because the points of the scissors reach into cavities that we cannot see and are not likely to slip, cut, or pierce inadvertently. The best kitchen scissors for this type of task have a serrated blade on one side, to grip the slippery fins and cut easily.

scaling & trimming

If you plan to eat the skin, then it is best to scale the fish. If, on the other hand, you are going to remove the skin before serving the fish, then there is no need to scale it. A salmon is shown here.

I Lay the fish on a work surface covered with a plastic bag or newspaper. If the fish is small, you can lay it in the bottom of the sink under running cold water. Take hold of the fish by its tail, then begin to scrape off the scales from the top side using a fish scaler. Scrape from the tail towards the head. Turn the fish over and scrape off the scales on the other side.

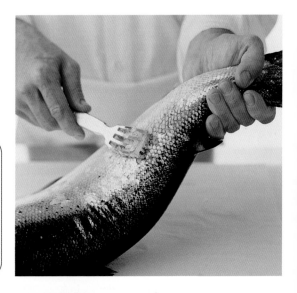

tools of the trade

If you don't possess a fish scaler, use a chef's knife to scale the fish – scrape off the scales with the spine of the knife blade.

2 Cut off the back (dorsal) fins and belly fins with kitchen scissors, then trim off the fins on either side of the head. If desired, trim the tail with the scissors to neaten it, perhaps cutting into a "V" shape.

boning through the stomach

To bone a whole fish in this way, it is first gutted through the stomach (*p102*), and then scaled and the fins trimmed. Once boned, it can be stuffed for cooking, mostly by baking. A sea bass is shown here.

1 Open up the fish. Loosen the ribcage (transverse bones) from the flesh on the top side by sliding a sharp knife (such as a filleting knife) along the ribcage. Turn the fish over and repeat, to loosen the transverse bones from the flesh on that side.

2 Snip the backbone at head and tail ends using kitchen scissors. Then, starting at the tail, peel it away from the flesh. The transverse bones will come away with the backbone.

boning from the back

Boning a whole round fish from the back prepares it for stuffing and baking. First scale the fish, then trim off the fins (p104). Do not gut the fish. Use a filleting knife, or other sharp, flexible knife, for boning.

1 **Cut down the back of the fish, cutting along one side of the backbone from head to tail. Continue cutting into the fish, keeping the knife close on top of the bones. When you reach the belly, don't cut through the skin.**

2 **Turn the fish over and cut down the back from tail to head along the other side of the backbone. Continue cutting as before, to cut away the flesh from that side of the backbone.**

3 **Using kitchen scissors, snip the backbone at the head and tail ends, then remove it. Pull out the guts (viscera) and discard. Rinse the cavity under cold running water and pat dry.**

4 **Pull out any pin bones (the line of tiny bones down each side of the fish) using large tweezers or small needle-nose pliers. The fish shown is black sea bass.**

quick tip

Sea bass is delicious baked whole with a tasty stuffing. It is not difficult to bone from the back, and the large empty cavity takes a filling exceptionally well. Before you begin boning a sea bass, use poultry shears to cut off the dorsal spines next to the skin, which are particularly sharp and unpleasant.

filleting

A round fish (red mullet is shown here) is typically cut into two fillets after it has been gutted. It is best to use a filleting knife, because the blade is long and more flexible than that of a regular kitchen knife.

1 Depending on the fish and whether you are going to leave on the skin, scale the fish (p104). Using a filleting knife, cut into the head end, just behind the gills, cutting with the knife at an angle just until you reach the backbone.

2 Starting near the gills, cut the fish down the length of the back, cutting along the top side of the backbone.

3 Working again from head to tail, continue cutting over the bone, keeping the knife flat and folding the fillet back as you cut. When the fillet has been freed, remove it.

4 Turn the fish over and repeat the process to remove the second fillet, this time cutting from the tail to the head.

top technique

Sometimes the skin causes the fillet to curl in the heat of the pan or grill. To prevent this, before cooking, score 4–6 shallow lines across the skin not quite to the sides of the fillet, using the point of a filleting or paring knife. Scoring also helps to transfer the heat quickly because the skin can otherwise act as a barrier.

skinning a fillet

If you plan to skin fish fillets, there is no need to scale them or the whole fish from which the fillets are cut, unless you want to fry the skin later for use as a garnish. Round-fish and flat-fish fillets are skinned in the same way. A whole salmon fillet is shown here.

1 **With the fillet skin-side down, insert a filleting knife into the flesh near the tail end, turning the blade at a slight angle. Cut through the flesh just to the skin.**

2 **Turn the blade of the knife almost flat and take tight hold of the end of the skin. Holding the knife firmly in place, close to the skin, pull the skin away so as to cut off the fillet.**

cutting salmon steaks

Any large round fish can be cut into steaks, but those most commonly found in steak form are varieties of tuna, swordfish, and salmon.

1 **Gut the fish (here a salmon) through the stomach (p102). Scale it, then trim off the fins (p104). Using a chef's knife, cut off the head just behind the gills.**

2 **Holding the fish firmly on its side, cut across to get steaks of the desired thickness.**

filleting a monkfish tail

A monkfish tail section is usually bought already skinned. However, if the tough black skin is still on, remove it with the help of a sharp knife. At the wider end of the tail, slide the knife under the skin, then take firm hold of the skin and pull and cut it away from the flesh.

I For this task, use a freshly honed filleting knife or utility knife. Cut down one side of the central bone to release the fillet on that side. Repeat on the other side of the bone to remove the other fillet.

2 Before cooking the fillets you need to remove the thin, purplish membrane that covers their skinned sides. Do this by sliding the knife between the membrane and the flesh, tugging away the membrane and cutting it off in small strips.

slicing gravlax & other sugar-cured fish

For this Swedish speciality, raw salmon is cured in a sweet salt mixture. Dill is the most common flavouring, although peppercorns or slices of orange or lemon are also sometimes used. The salmon is sliced paper-thin for serving.

slicing **Rinse off the cure and pat the fish dry with kitchen paper. Lay the fillet out flat, flesh-side up, and slice very thinly on the diagonal, cutting away from the skin with a paring knife (shown here) or granton knife. Gently lift the slices off the skin and serve.**

skinning & gutting an eel

Eel is easier to skin immediately after it has been killed, and it should be cooked as soon as possible after skinning and gutting. In general, eels will come to you already killed, but with skin intact. The skin is very slippery so you'll need to use a towel to help you grip it.

1 Using a kitchen towel, hold the eel near its head. With a large chef's knife, cut the skin all around the base of the head, not cutting completely through.

2 Take hold of the head with the towel, and use pliers to pull the skin away from the cut made around the base of the head.

3 Still grasping the head end with a towel in one hand, take hold of the freed skin with another towel in your other hand. Pull firmly to peel off the whole skin.

4 Starting at the head end, use kitchen scissors to make a cut down the length of the underside of the eel. Remove the guts (viscera). Rinse the eel in cold running water and pat dry.

top technique

The nervous system of a freshly killed eel can keep it surprisingly agile, sometimes enough to twist itself round your arm. You might find it easier to hang the eel by its head on a strong hook, cut the skin around the head (see step 1), then peel the skin off using pliers. Once skinned, cut the eel into sections for cooking.

serving whole cooked fish

After cooking a whole fish, the easiest way to serve it is to transfer it from its baking dish to a cutting board to prepare while still in the kitchen, and then to a platter to serve. Whole round fish are easily served using a fork, large spoon, and table knife or fish server.

I Carefully peel away the skin from the top of the fish (red snapper is shown here), cutting it from the head and tail if these are left on. Scrape away any dark flesh, and scrape off the bones that lie along the back of the fish.

2 Cut down the centre of the fish with the back of the spoon and a table knife, then lift off the top 2 fillets, one at a time. Snap the backbone at the head and tail ends, and lift it out. Replace the top fillets to reshape the fish.

RAW FISH FILLETS

There are many ways to serve raw seafood, the most popular being the Japanese specialities, sushi and sashimi. Almost any type of fish and many shellfish can be used for sashimi and sushi, as long as the seafood is perfectly fresh. Seafood to be sliced for sushi or sashimi should be frozen first, for about 30 minutes. This will make it easier to slice very thinly.

slicing raw fish **The fish can be cut into any shape in order to give the presentation you want. Here, a yellowtail tuna fillet is trimmed into a block, so that thin, square slices can be cut, but purists simply cut along the shape of the fish.**

With a long-bladed knife, such as a Japanese *hancho hocho* or a freshly honed utility knife, cut the fish into very thin slices – about 3mm (⅛in) thick.

dicing raw fish **Trim the tuna to make a neat block, using a Japanese *tako hiki* or *yanagi ba* or a freshly honed utility knife. Then cut it into slices about 5mm (¼in) thick. Cut each slice into strips the same width as the thickness, then cut across the strips to make dice.**

apply this skill

The most common fish used for sushi and sashimi include clams, mackerel, octopus, sea urchin roe, salmon, squid, tuna, and hamachi (yellowtail tuna). Unagi (eel), which is cooked, is also popular in sashimi and sushi.

the **sliced** sashimi is artfully presented with a **garnish** and **dipping** sauce

FLAT FISH

Flat fish, a type of "fin fish", are flat and oval-shaped, with eyes on one side of the body. They typically have coloured top sides, which may be dark brown, black, or dark grey and white undersides. The most common preparation techniques are gutting and trimming, skinning, boning, and filleting.

gutting & trimming

If you plan to serve a flat fish whole, this is the first part of the preparation. Flat fish are normally gutted first to ensure there are no viscera to cut into when the fish is being trimmed. Then the fins are trimmed and the fish is scaled, if necessary. A plaice is shown here.

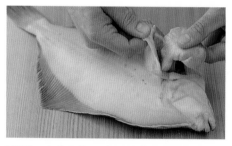

1 With a chef's knife, make a small cut along the stomach so you can reach in to remove the guts (viscera) and any roe. Discard these.

2 Use kitchen scissors to trim away the fins. Leave about 5mm (¼in) of fin still attached to ensure that you don't cut into the fish body when trimming.

3 Scale the white side, if necessary (see opposite), then cut off the gills with scissors and discard them. Rinse the fish inside and out under cold running water.

top technique

To serve a flat fish whole but without the head, use this easier way to gut it. After trimming and scaling, lay the fish dark-side up; make a V-shaped cut around the head. Grasp the head gently but firmly and, with a quick twisting turn, pull the head away. The guts (viscera) and the gills should come out with the head. Rinse.

scaling

If the skin on the white side of a flat fish feels rough to the touch, scale it after gutting the fish and trimming off the fins (see *opposite*). Lay the fish on newspaper or a plastic bag.

> ## top technique
>
> If you don't have a scaler, use the back of a knife. Grasp the fish by the tail and, at right angles to the skin, rub hard, ideally under running water.

removing scales **Using a fish scaler, scrape off the scales, working from the tail towards the head. The dark side isn't scaled because this skin will be removed before serving.**

skinning

This is the second part of the sequence if you want to cook a flat fish whole, either on the bone (see *opposite*) or boned (*p117*) and perhaps stuffed. Only the dark skin is removed – it is tough. The white skin is left on to help retain the fish shape during cooking.

1 Using kitchen scissors, trim the fins from the belly and back, leaving about 5mm (¼in) of fin still attached to the fish (here a turbot). Turn the fish white-side up. Make a small cut at the tail end to separate the dark skin from the flesh.

2 Insert a utility knife between the flesh and the dark skin. Keeping the knife blade flat against the skin, take hold of the skin at the tail firmly with your other hand and pull the skin away to cut off the flesh neatly.

skinning & filleting a Dover sole

Dover sole requires special handling, differing from the preparation of other flat fish. Most chefs prefer to skin Dover sole prior to filleting; however, if the sole is being prepared to cook whole, the skin is left on. Note that only the black skin is removed. The delicate white skin remains intact, even when the fish is cut into fillets.

quick tip

To get a good grip on the skin when pulling it from a fish, you can either grasp the flap of skin in a towel or dip your fingers in salt first. Pull off the skin sharply, parallel to the flesh and as quickly as possible.

1 Make a small cut with a paring knife through the skin at the tail end, cutting at an angle, to separate a flap of the dark skin from the flesh.

2 Using a towel, take hold of the freed flap of dark skin securely. Holding the tail end of the fish firmly with your other hand, pull the dark skin away from the fish. Fillet the fish to make 2 fillets (p118).

boning

Flat fish to be cooked whole with a stuffing should have the bones removed. Prepared like this, a flat fish makes a beautiful presentation. A turbot is shown here.

Skin the fish (*see opposite*). Lay it skinned-side up on the board and, using a filleting knife, make a cut down the centre, cutting through the flesh just to the backbone.

Free the fillet on one side from the bones by cutting horizontally to the outer edge of the fish. Do not remove the fillet. Turn the fish around and repeat to free the other fillet.

Slide the blade of the knife under the backbone, down the length of the fish, to loosen the bone from the flesh (*see left*).

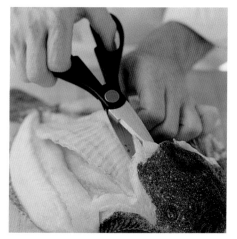

2 Use kitchen scissors to snip the backbone at the head and tail ends of the fish, as well as in the centre, to cut it into pieces.

3 Carefully lift the pieces of backbone from the fish, cutting them from the flesh with the knife where necessary. Before stuffing the fish, check to be sure there are no bits of bone.

cutting two fillets

Flat fish tend to be wider in span than round fish and therefore can be cut into either two or four fillets. The choice is usually governed by the size of the fish (here a flounder) as well as by how it is to be served.

1 Gut the fish (p114), trim the fins, and cut off the gills. Lay the fish with its head end nearest to you. Cut down to the backbone at the base of the head. Insert the filleting knife, starting at the tail end, into the outer edge of the fish, cutting just above the bones.

2 Continue cutting around the edge towards the head. Turn the fish around (not over) so the tail is nearest to you. Starting at the head end, cut along the outer edge on the other side, again cutting just above the bones and continuing towards the tail.

3 Turn the fish around. Keep the knife blade almost horizontal and close to the bones, and cut gently with long, smooth strokes. Continue cutting over the centre ridge of bones and towards the other side.

4 Carefully lift off the top fillet in a single piece. Turn the fish over and repeat the process on the other side to free the second fillet, this time starting the cutting (see step 3) at the head end. Skin both fillets.

cutting four fillets

This sequence shows how to fillet a very large flat fish (turbot is shown here) into four "quarter fillets". Before filleting, gut the fish by taking off the head (*p114*); this will remove the gills too. Then trim the fins.

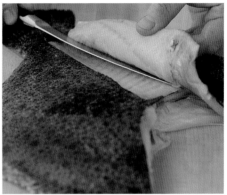

1 Lay the fish flat on a chopping board with the tail end nearest to you. Make a cut down the centre of the fish with a filleting knife, following the visible line that separates the fillets and cutting through the flesh just to the backbone.

2 Turn the blade of the knife flat against the backbone on one side and, with long, smooth strokes, cut horizontally until you reach the outer edge. Cut through the skin at the edge and remove the fillet. Repeat for the other fillet on the same side.

3 Turn the fish over. Following the same process, cut off and remove the 2 fillets on that side of the fish. Skin all 4 fillets.

quick tip

Not all flat fish are handled in the same way. For instance, while most are cut into fillets and then skinned, a Dover sole is skinned first (*p116*).

filleting a skate wing

Skate is a type of ray fish. It is prized for its "wings", which are sometimes filleted to remove the meat from the gelatinous cartilage. Filleted skate wings are most commonly just lightly sautéed, although they are also delicious poached in a light court bouillon.

Lay the skate wing on a board with dark-side uppermost and the thickest side nearest to you. Using a filleting knife, cut into the flesh on the thickest side until you reach the cartilage, which is about halfway down.

2 Turn the wing around. Turn the knife flat on the cartilage and cut the flesh away until you reach the outer edge of the wing. Cut along the edge and detach this fillet. Repeat on the other side. Remove the skin from the fillets as for round fish (p108).

serving whole flat fish

For large flat fish such as Dover sole, as shown here, you can use a table knife and large spoon for serving at the table. Place the fish on a hot serving platter.

With the knife and spoon, push away the fin bones from both sides of the fish. With the edge of the spoon, cut along both sides of the backbone, just cutting through the flesh to the bone. Lift off the top 2 fillets, one at a time.

2 Lift out the backbone and set it aside to discard. You can replace the top fillets for an attractive presentation.

SHELLFISH

The shellfish family is made up of crustaceans and molluscs. Crustaceans, such as lobsters, prawns, and crabs, have an exterior skeleton, segmented bodies, and jointed limbs. Most molluscs have one or two hard shells, except for octopus and squid, which don't have a shell.

opening oysters

To open oysters, use a thick towel or napkin or wear a special wire mesh glove to protect your hand from the sharp edges of the shell. If you intend to serve oysters raw, scrub them well before opening.

fresh oyster **Insert an oyster knife into the crevice at the point of the shells. Push gently to sever the muscle hinging the shells together; pushing the knife too deep will damage the oyster. Twist the knife, to prise the shells apart, then separate them carefully with your fingers, keeping the bottom shell level, so the liquor does not spill. With a teaspoon, scrape the oyster off the flat shell and transfer it to the rounded shell. The liquor should be clear and briny; if it is cloudy, you have pierced the oyster. Discard the tough sinew. Serve the oysters on ice.**

quick tip

When buying oysters and clams, check that the shells are tightly closed. Discard any with broken shells, as well as oysters that smell "fishy" on opening. After boiling or steaming molluscs such as clams and mussels, discard any that are still closed.

opening clams

All clams should be scrubbed well and, as wild clams tend to be very sandy, they may need to be "purged" – put in a large bowl of cold water with some cornmeal or polenta and leave to soak overnight in the refrigerator. Then they can be opened and eaten raw or cooked. Alternatively, they can be boiled or steamed to open the shells.

1 Holding the clam in a thick towel to protect your fingers, work the tip of the clam knife between the top and bottom shells, then twist the knife upwards to force the shells apart.

2 Slide the knife over the inside of the top shell to sever the muscle and release the clam, then do the same to release it from the bottom shell. Take care not to cut into the meat. To serve raw on the half shell, snap off the top shell. For soft-shell clams, remove the dark membrane before serving.

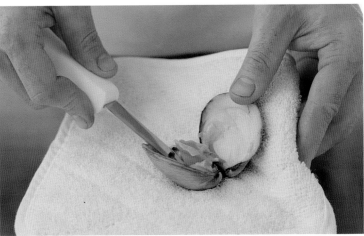

opening scallops

Although most home cooks will buy scallops already opened, you
will sometimes find them in the shell. Scrub the shells clean before
opening. The scallops can then be served raw or cooked.

1 Holding the scallop firmly in your hand, flat shell uppermost,
insert a long, thin, flexible knife in between the top and
bottom shells, keeping the blade as close to the inside of the
top shell as possible to avoid damaging the scallop meat inside.
Slide the knife around the top shell to sever the muscle.

top technique

To ensure the scallops don't
nip you, place them over a
gentle heat. When the shells
have separated, by not more
than 1cm (½in), hold the
scallop in your palm. Insert a
table knife between the shells
and scrape everything off
the flat shell. Keep the blade
angled slightly down, towards
the flat shell, so it does not
damage the scallop meat.

2 When the
scallop meat
has been freed
from the top shell,
remove the shell.
Detach the scallop
from the bottom
shell with the help
of the knife, again
taking care not
to cut into the
scallop meat.

To prepare abalone, you need to cut with the point of a paring knife around the inside of the shell to free the foot, which then needs to be trimmed of any dark skin, fringe, and viscera. Abalone is often eaten raw. Alternatively, slice it thinly and sauté quickly – abalone should be cooked briefly as otherwise it will be tough.

3 Pull or cut away the viscera and fringelike membrane from the white scallop and coral; discard the viscera and membrane. Rinse the scallop and coral before use.

peeling & deveining prawns

Prawns contain a small sand line, also known as the intestinal vein.
Unless the prawns are small, the vein is usually removed before cooking.
This is done because the vein is gritty on the palate.

1 Pull off the head, then peel off the shell and legs with your fingers. Sometimes the last tail section is left on the prawn. Save heads and shells for use in stock, if desired.

2 Run a paring knife lightly along the back of the prawn to expose the dark intestinal vein. Remove the vein with the tip of the knife or your fingers. Rinse the prawn under cold running water and pat dry.

butterflying prawns

open for stuffing **Make a cut along the back so that the peeled prawn can be opened flat, like a book. Do not cut all the way through. Remove the vein with a paring knife, then rinse the prawn and pat dry.**

flattening prawns

large prawns **Lay the prawn so that the inside faces you. Make 6 nicks in it with a sharp knife (here with a Japanese knife) to stop it curling as it cooks. Using the side of a knife, flatten the prawn to expel any water.**

tools of the trade

To open the tail shells of langoustines, use a pair of fine-pointed scissors. With the belly facing you, snip up the length of each flat, transparent shell to release the soft tail meat.

langoustines are also called **Dublin Bay prawns** or **scampi**

cleaning a live blue crab

The cleaning process described here will prepare a hard-shell blue crab for cooking in a soup or sauce. Alternatively, it can be done after the crab has been boiled or steamed (skip step 1).

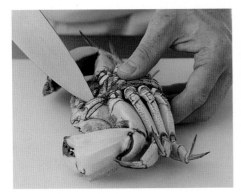

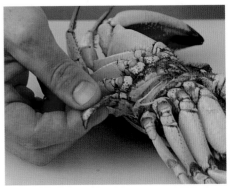

1 Hold the crab on its back on a chopping board. Insert the tip of a chef's knife into the crab, directly behind the eyes, and quickly bring the knife blade down to the board to kill it.

2 Pull and twist off the small, folded tail flap (the apron) from the underside of the crab. Female crabs have rounded aprons; male crabs have thin, pointed aprons.

3 Press down on the centre and leg section of the crab, and pull off the top shell.

4 With kitchen scissors, snip off the gills (dead man's fingers). Discard the spongy sand bag that is located behind the eyes.

5 Cut the crab into halves or into quarters. It is now ready for cooking.

cleaning a live soft-shell crab

Soft-shell crabs are blue crabs that have moulted their hard shells. Popular ways to cook them are deep-frying and sautéing. The entire crab is eaten – the newly formed shell is crunchy and delicious.

1 | With kitchen scissors, cut across the front of the crab to remove the eyes and mouth and kill it.

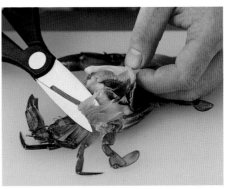

2 Fold back the top shell so you can snip away the gills from both sides.

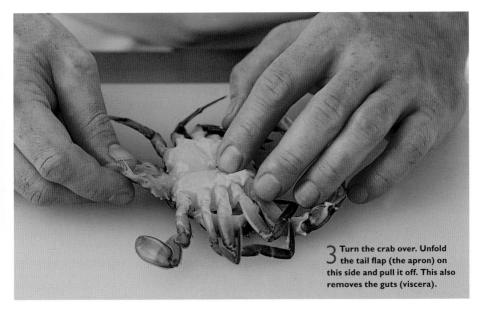

3 Turn the crab over. Unfold the tail flap (the apron) on this side and pull it off. This also removes the guts (viscera).

removing the meat from a cooked crab

All large, meaty crabs have claws, legs, and body. Shown here is a common European crab, which contains soft brown meat as well as white meat. Dungeness crab is prepared in much the same way.

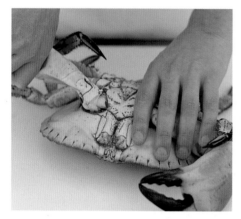

I Set the crab on its back on a chopping board and firmly twist the claws and all the legs to break them from the body.

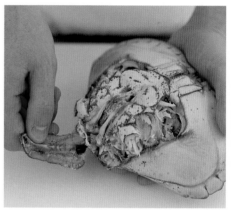

2 Lift up the tail flap or apron (here a triangular male flap) on the underside of the body, then twist it off with your hand and discard.

3 Crack the central section of the shell under the tail. Using your thumbs to start off, prise it apart, then lift off the shell. Remove any white meat from the shell using a teaspoon.

4 Pull off the gills (dead man's fingers) from the sides of the central body section and discard them. Also discard the intestines, which will either be on each side of the shell or be clinging to the body.

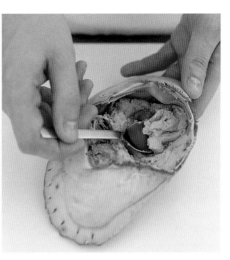

5 Use a large chef's knife to crack or cut the central body section into several large pieces. Dig out the white meat using a lobster pick or skewer, discarding any membrane. Reserve the white meat in a bowl.

6 Spoon out the soft brown meat from the shell and reserve it to serve with the white meat (there is no brown meat in a Dungeness crab). Discard the head sac. If there is any roe, spoon this out too and reserve it.

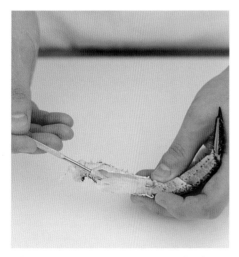

7 With poultry shears or the spine of a chef's knife or santoku knife, tap one side of the shell on each leg to crack it. Lift out the meat, in 1 piece if possible, using a lobster pick. Add to the white meat from the body.

8 Crack the claws with special lobster crackers, a nutcracker, or a small hammer and extract the meat. Check all the white meat for bits of membrane and shell before serving.

cleaning a lobster

To clean and cut up a live lobster before cooking, reserve the tomalley (greenish liver) and coral (the roe, which will be black) to use in a sauce, butter, or stuffing. The head, body, and legs can be used in a fish stock.

1 Leave the rubber bands in place around the claws. Lay the lobster flat on a cutting board and hold it firmly. Put the tip of a heavy chef's knife into the lobster's head to kill it, then cut straight down and split it in 2.

2 Remove the claws by twisting them off the lobster or, if necessary, by cutting them off with the chef's knife.

3 Take hold of the body and head section with one hand and the tail section with the other hand. Twist to separate them.

4 Spoon the tomalley (liver) and any coral from the head and tail sections, and reserve. The tail section and claws are now ready for cooking.

splitting a lobster

To cut a lobster in half you need a large heavy chef's knife and a bit of elbow grease. Hold the lobster firmly as you cut it. The halves can then be grilled as they are or used for lobster thermidor.

1 Hold the lobster parallel to the edge of the chopping board and kill it (*see* step 1, *opposite*). The lobster won't be sentient at this stage, but do expect some twitching until you finally cook it.

2 Turn the lobster round, quickly flatten its tail, and pin to the board with one hand on its back. Then, from the point of the first incision, draw the knife downwards to cut the lobster in half.

3 With a teaspoon, take out the coral and tomalley (*see right-hand bowl*) and use in another recipe. Discard anything greeny-beige. Cook the cleaned lobster (*see foreground*) in the shell for juicier flesh.

apply this skill

The tails and claws (*see opposite*) are ideal for simmering in court-bouillon and serving cold with mayonnaise, or steaming with herbs and serving in broths. Crack the claws before serving them.

Cook a split lobster (*see left*) cut-side down in a hot pan or char-grill and serve with lemon wedges. Alternatively, turn them over and finish under a hot grill, topped with a creamy Parmesan sauce. The eggs may also be used.

taking the meat from a cooked lobster

Rather than cut up a live lobster before cooking (pp132–33), you can cook the lobster whole and then take out the meat.

I Take firm hold of the tail section and twist sharply to separate it from the body and head section.

2 Turn the tail section of the lobster over and, using kitchen scissors, cut down the centre of the flat underside of the shell.

3 With your thumbs, press on both sides of the cut and pull open the tail shell. Remove the meat in one piece.

4 With lobster crackers or the spine of a santoku knife, crack open the claw shells. Take care not to crush the meat inside.

5 Remove the meat from the claws, in whole pieces if possible. Discard any membrane attached to the meat.

quick tip

If the lobster is large, you may find it easier to remove the tail meat by snipping down each side of it, right next to the pink tail shell. The white, soft under-skin will peel off intact, and the large, fat tail come out whole and ready for cutting up.

cleaning squid

Squid is made up of the mantle (body) and tentacles (the "arms").
It has one eye and a plastic-like inner lining (called the quill) in the
mantle. The eye and the quill must be removed before cooking.
Perhaps the most interesting part is its mouth, which is referred to
as the "beak". This is a small ball shape in the middle of the tentacles.

1 Pull the mantle and the tentacles apart. The eye, viscera, and ink sac will come away with the tentacles, attached to the head.

2 Pull the transparent, plastic-like quill out of the mantle and discard.

3 With a small chef's knife or utility knife, cut the tentacles from the head, cutting just above the eye. Discard the head and the viscera, but keep the ink sac if you want to use the ink to colour a sauce.

4 Open the tentacles and pull out the beak. Discard it. Rinse the mantle and tentacles under cold water. Leave the mantle whole or cut it into rings, according to the recipe; leave the tentacles whole.

scoring mantle of squid

Instead of cutting squid mantle into rings, cut it into strips for stir-frying. Carefully scoring it first will make it roll up when fried, producing tight, juicy curls of squid for the finished dish.

I You will find a creamy line running down the squid, where the quill was attached, inside. Using this as a guide, gently slit through one side with a little Japanese ceramic knife. Open out the mantle and place the inner side flat and upwards on the chopping board. Scrape the mantle to discard any excess flesh and make it as thin as possible.

Hold the knife delicately, so you cut only to a one-third depth, and score the entire mantle with lines that are 1.5cm (½in) apart, in a criss-cross pattern. (Scoring works best on small squid.)

2 Cut the mantle into pieces about 4cm (1½in) square (shown here) or into triangles. Lay the pieces cut-side up in a little smoking-hot oil and cook with some chilli dice until they curl, or char-grill.

quick tip

Cuttlefish is prepared in a manner similar to squid; the main difference being that you cut the side to remove the innards and ink sac. The mantle is always cut into slices rather than rings.

cleaning & sectioning octopus

With a small octopus all you need to do is cut the head away and section the meat. A larger octopus (over 4.5g/10lb) needs to be be tenderized by pounding with a kitchen mallet for about five minutes.

1 Using a sharp chef's knife, cut the head away from the octopus; discard the head. Cut away the eyes.

2 Turn the octopus over and, using a small utility knife, cut out the beak (mouth) from the underside. Discard the beak.

3 Slice off the tentacles with a large utility knife, then cut them into sections of the desired size.

cleaning sea urchins

Sea urchin roe is creamy and tasty. It is commonly eaten raw, most often as sushi, but may also be lightly puréed for use in sauces, or poached. It tends to be at its best in the colder months.

1 Hold the sea urchin, top-side up, in a thick towel (the top has a small hole opening in the centre). Insert the point of kitchen scissors into the opening, then cut out a lid.

2 Remove the cut-out lid to expose the roe. Spoon out the sacs of roe, taking care as they are very delicate.

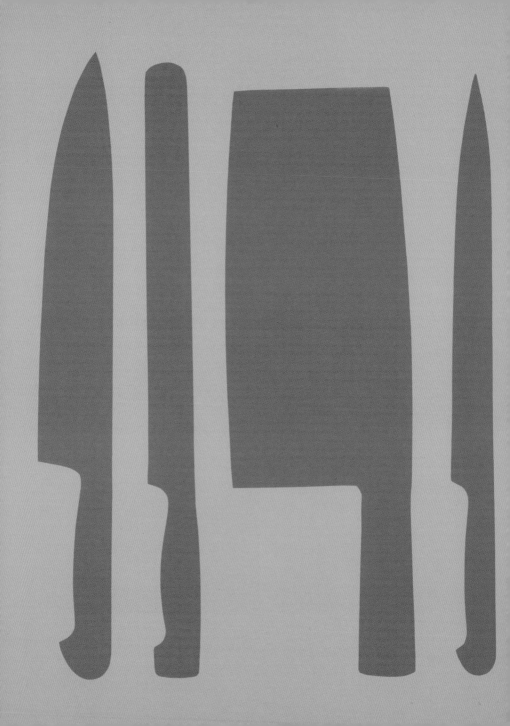

MEAT, POULTRY & GAME

MEAT, POULTRY & GAME

Understanding the anatomy of the skeleton, muscles, and ligaments makes meat, poultry, and game easier to cut and carve.

anatomy of meat

Animals have a strong and complex skeleton to support a heavy frame of long muscles. Muscle fibres, often over 30cm (12in) long, can be easily discerned as a coarse grain, in tough meat from mature animals and in continuously used muscles such as the neck and forelegs. Connective tissue sheaths the muscles in blue-white membranes; sometimes visible as a transparent silver skin, for example surrounding beef fillet, sometimes as a hefty sheet of gristle, as in brisket. At the end of the muscle, the connective tissue becomes a shiny, white tendon, anchoring the muscle to the bone. When boning a leg of lamb, you will see a tendon in the centre of the ball and socket joint.

Use gravity to your advantage when boning meat or slicing off gossamer connective tissue, such as from a fillet of pork: hold up the bones and let the meat and fat fall to the work surface, giving you a better view so you can cut precisely and avoid cutting into the muscle. Don't slice away too much gristle – the meat will simply fall apart. Instead use a slow cooking method, as in braising oxtail, and the gristle will dissolve.

Poultry and game birds have a featherweight skeleton and light and dark muscles: muscles that are exercised more frequently store more oxygen and provide dark meat. Chickens and turkeys do little, if any, flying and have tender, white breast meat, or wing muscles, but darker legs. Game birds such as wild duck, on the other hand, consist only of dark meat and have much stronger leg tendons.

tools for cutting & carving

The challenge of preparing meat, poultry, and game for the table requires a range of cutting tools. These include a cleaver and boning knife for heavy work; a long, fine knife and meat fork for carving and slicing; scissors and scalpel for crackling or pork skin; small knives for scoring fat; a granton knife for cold cuts; and even an electric slicer for paper-thin slices of ham. Poultry shears are needed to trim feathers and breast bones, and paring knives to cut birds.

If you would like to "break down" a carcass into smaller cuts and joints yourself, get a real butcher's block, which is much lower than a normal work surface, and a scimitar.

Cooking meat on the bone gives you a chance to carve it with skill and a flourish at the dining table, to the admiration of your guests.

BEEF

Beef is cut from the carcass of our largest farmed animal, and yields tough and tender meat, all of which can be delicious, depending on how we butcher, cut, cook, and carve it.

cutting across the grain

Braises, casseroles, stews, and other slow-cooked dishes demand tough meat, full of flavour. To enjoy them, the meat must be cut across the grain and cubed – it would be very stringy and unpleasant otherwise.

slicing **Trim the fat, using a straight-edged slicing knife. The "grain" of the meat is formed from developed muscle fibres. This is recognized easily in cuts of tough meat with a coarse grain such as brisket (shown here), chuck, and topside. Slice the meat first, then cut into cubes or rectangles, working at right angles to the direction of the muscle fibres.**

cutting with the grain

The fillet is the least-used muscle in the beef carcass. It is recognizable by its satiny texture. Aged fillet is so tender that you do not have to cut across the grain.

cutting strips **Chill the fillet and, if it is thick and round, cut it into half horizontally. Place the flat side against the board. With the tip and point of a utility knife, cut carefully into even strips, following the direction of the muscle fibres.**

cutting fillet steak for sautéing

Prime, lean meat is best for this quick cooking method in which cubes or strips of meat are tossed frequently so that they "jump" in the pan. To get suitably even-sized pieces of meat, it is best to cut it yourself.

cutting cubes **Cut the fillet crossways into 7.5cm (3in) pieces, then cut each piece into slices against the grain, about 7.5cm (3in) square and 2.5cm (1in) thick. You could use a scimitar (shown here) or a small slicing knife. Lay the slices flat and then cut the squares into 7.5x2.5cm (3x1in) strips.**

carving entrecôte steaks after grilling

Prime, tender cuts of meat like entrecôte, or sirloin, steak are also suitable for grilling, which is either cooking under the grill, on the grid of a barbecue, or on a ridged cast-iron grill pan on top of the stove.

carving **After grilling, allow the steaks to rest, then use a small slicing knife to slice each steak diagonally against the grain into 4 or 5 thick slices, using a gentle sawing action with the knife.**

carving roast rib of beef

After roasting, transfer the meat to a carving board, cover loosely with foil, and leave to rest in a warm place for 15–30 minutes before carving. Skilful carving makes all the difference to the taste and texture of a roasted meat.

1 To carve, stand the roast with the ends of the bones facing up. Steadying the meat with a carving fork on the fatty side, and using a sawing action with the slicing knife, cut downwards between the bones and the meat to separate them.

2 Discard the bones and put the chunk of meat, fat-side up, on the board. Cut downwards across the grain into thin slices, again using a sawing action with the knife.

quick tip

Letting the meat rest for at least 15–30 minutes after roasting allows the muscles to relax so the juices are retained within the meat and carving is easier. The meat will not go cold during this time – as long as it is not cut into, it will stay hot inside.

serve the slices of beef on warmed plates to ensure the meat doesn't go cold

LAMB

The best-quality meat will come from the back half of an animal, in particular the fleshy hindquarters. The toughest cuts come from those parts of an animal that move the most.

boning a saddle of lamb

This classic technique of boning a saddle of lamb from underneath keeps the saddle whole – and gives a beautiful meaty joint that is perfect for stuffing. First, detach any kidneys and reserve for stuffing.

1 Strip off the membrane covering the fatty side of the saddle. Turn over and remove the 2 fillets from either side of the backbone, or chine. Work from one side of the bone outwards, stroking against it with a flexible boning knife until the fillet is released. Repeat on the other side. Reserve the fillets.

2 With the tip of your knife, loosen the outside edge of I end of the backbone, then cut around and underneath it on I side, working towards the middle. Continue stroking with your knife along its length, gently lifting up the bone to help release it from the meat.

3 Repeat the stroking and lifting technique on the other side and then work the knife under the backbone so that it comes free, taking great care not to pierce through the skin. The 2 "canons" of meat underneath will now be revealed.

4 Clean the meat and fat from the flaps by stroking with your knife from the middle outwards – keep going until you get to the fat on the skin and the flaps are smooth. Square off the edges, making them about 12cm (5in) from the canons. Turn the joint over and very lightly score through the fat and skin with the tip of the knife.

French cut of best end

Also known as *carré d'agneau*, this is one of the best roasts imaginable, with a crisp outer crust and sweet, juicy meat within. When preparing an untrimmed joint, aim to protect the single lean muscle, which runs next to the backbone, and don't cut it inadvertently off the ribs.

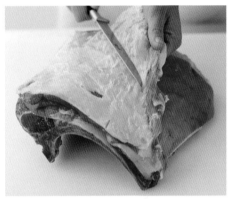

1 On one side of the joint, there is a cartilaginous and soft shoulderblade. Lift up the flap and cut it out with a flexible boning knife, holding the knife blade flat to avoid cutting into the muscle. Reserve it for stock.

2 Slip the boning knife beneath the membrane, which is often papery and crackly, and with your other hand, rip it off, exposing the fat. If the fat is too thick, shave off in layers with the boning knife.

3 With a small chef's or utility knife, cut between the ribs, running the blade as close as you can to the bones, to expose them and release the flap of fat – and a bit of meat – that holds them together.

4 Holding the knife blade about 4cm (1½in) away from the muscle, slice off the fatty flap. If you prefer really lean meat, cut closer to the muscle. Reserve the flap and trimmings to make stock.

5 Lay the rib bones on the chopping board. Take a cleaver (either an Asian one or a Western one as shown below) and, with one stroke, chop off the ends of the bones so that they all come off in a straight line. With the utility or chef's knife, carefully strip off the fine skin that surrounds the bones for an attractive presentation.

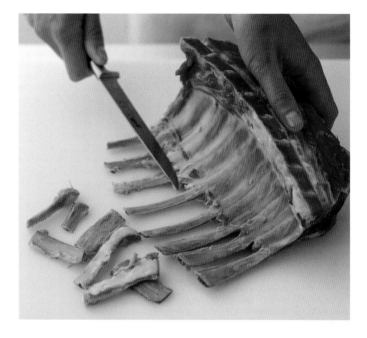

6 Hold the joint upright. With your cleaver, chop off the backbone, also called the chine, from the fattest end of the muscle. Try to do this with just 1 or 2 clean and accurate strokes. If you leave any traces of backbone on the joint, it will prevent you from slicing the cutlets apart after cooking.

quick tip

For barbecues, most cuts of lamb are excellent because they are small. Butterflied leg of lamb (*pp150–51*) will feed a crowd. Alternatively, cut the legs, best end, and loin into chops and cutlets. To avoid the outside of the joint charring and the inside being undercooked, marinate the meat first in yoghurt and herbs.

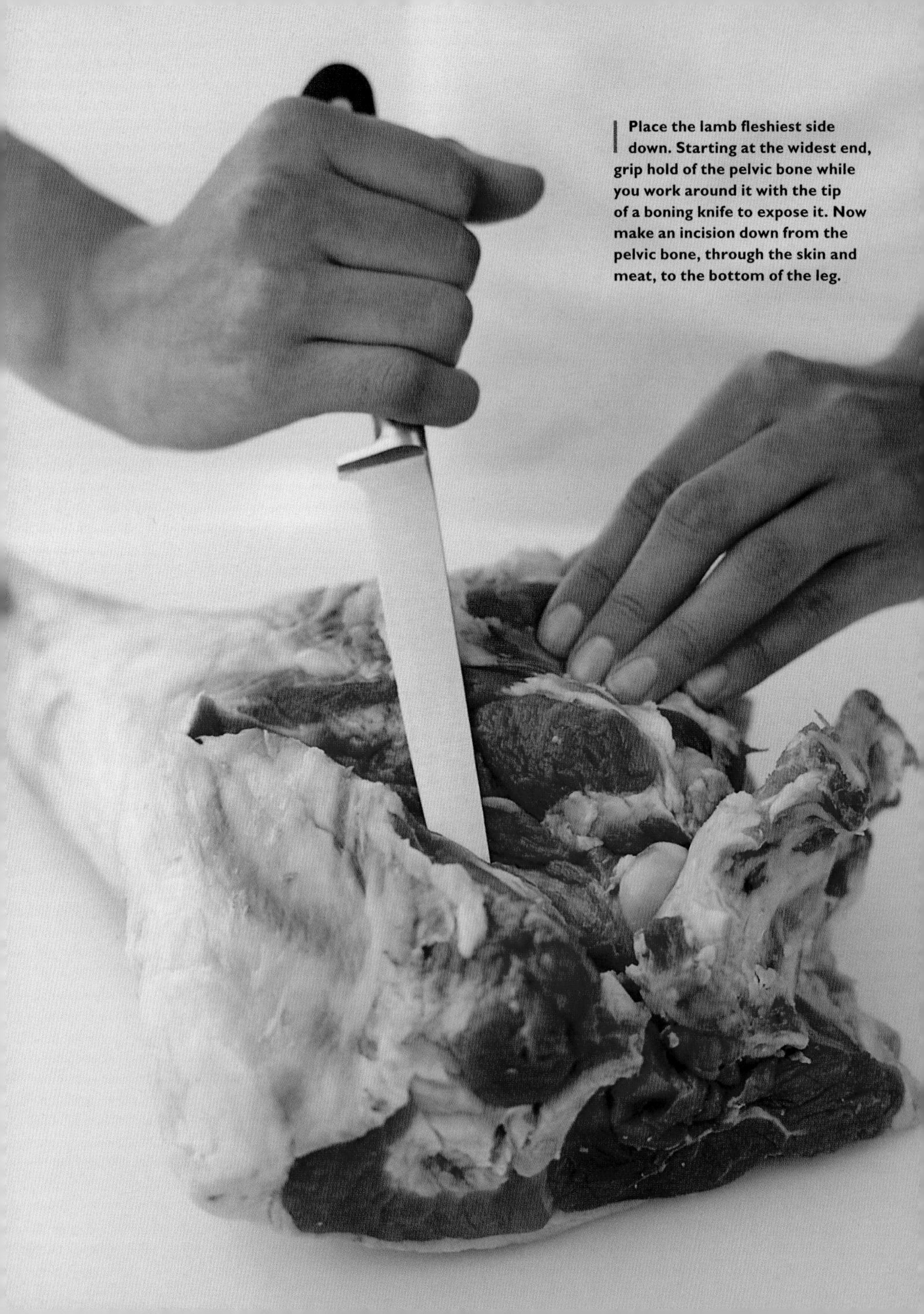

Place the lamb fleshiest side down. Starting at the widest end, grip hold of the pelvic bone while you work around it with the tip of a boning knife to expose it. Now make an incision down from the pelvic bone, through the skin and meat, to the bottom of the leg.

butterflying a leg of lamb

The secret of boning any piece of meat is to understand the anatomy. A leg is made up of three bones – pelvic, thigh, and shank. The pelvic bone is at the widest end, the shank at the narrowest. This technique is ideal for meat that is to be grilled or barbecued.

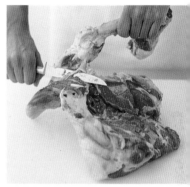

2 Work the knife around the thighbone, stroking all the way round close to the bone, to release it from the meat. Work with the tip of the blade, using short strokes to prevent tearing.

3 Using small stroking movements with your knife, always keeping the blade against the bone, keep going past the ball and socket joint and on down the length of the shank bone.

4 When you get to the bottom of the leg, cut through the sinew and tendons to release the end of the bone. You can then lift out all 3 bones (pelvic, thigh, and shank) in one piece.

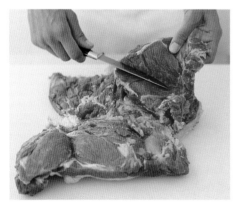

5 Open out the meat so that it lies flat of the knife, and make horizontal cuts with short stroking movements through the thick meaty "wings" on either side.

6 Open out the wings to reveal the boneless butterflied leg. If the meat is very uneven, cut slivers of flesh from the thickest parts and rearrange them in the thinner areas.

tunnel-boning a leg of lamb

This is a technique for the adventurous cook (or your butcher), but it is not at all difficult. Time and patience are all you need – and a freshly honed boning knife with a firm blade. Work slowly and consistently and you will be amply rewarded with a meaty whole joint with the bones "tunnelled out", which can be stuffed, wrapped, and carved with ease.

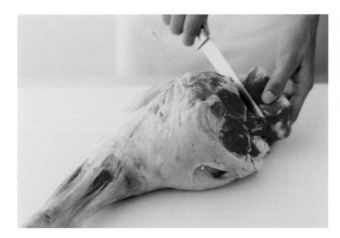

1 Place the lamb on a board with the fleshiest side face down. Starting at the top of the leg, locate the pelvic bone with your hand and grip hold of it while you work around it with the tip of your boning knife to expose it. Keep as close to the bone as possible.

quick tip

Feel free to roll the meat back and forth as you work (do whatever you feel most comfortable with), taking care not to tear the lobes of meat apart from each other. It is important to keep the joint as whole as possible, so that it does not fall to pieces when carved.

If the thighbone is proving difficult to sever from the ball and socket, either use a heavy cleaver or rest the joint on your thigh and snap the bone into two pieces.

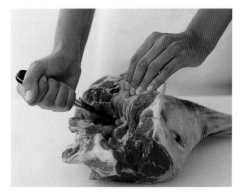

2 After the pelvic bone is exposed, slowly tunnel your way inside the leg, working all the way down the attached thighbone to the bottom. Keep picking up the meat and turning it over to get the best grip, taking care to keep the knife close to the bone so that you do not cut into the surrounding flesh.

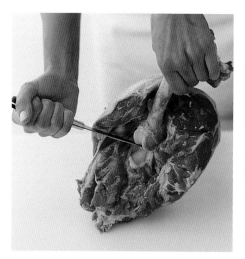

3 **When you get to the bottom of the thighbone where it joins the shank bone at the ball and socket joint,** grasp the end of the thighbone like a handle. Using the knife like a dagger, sever the ball and socket joint, at the same time pulling and wiggling the thighbone with your hand to work it free (see *box opposite*).

4 **Having taken the thighbone out, repeat the boning process with the shank bone in exactly** the same way until you get to the very bottom of the leg. At this stage it is very easy to see the "tunnel" that you have created.

5 **Stop working on the inside and** pull the shank bone through to the outside of the leg so you can get at it easily. Now continue working around the bone with your knife until it is released and the leg is completely bone free. You can use the bones to make stock.

stud with **garlic** and **rosemary** before cooking for a flavoursome roast

apply this skill

Studding a joint with rosemary and garlic imparts flavour to the meat. This technique can be applied to any joint. Tear the tops off the rosemary and halve each garlic clove lengthways. Make 12 incisions in the fat side of the lamb with the tip of a paring knife. Insert the rosemary and garlic into the slits in the meat.

carving a leg of lamb

Remove the lamb from the oven. Transfer to a carving board. Cover loosely with foil and leave to rest in a warm place for 15–30 minutes.

1 Holding the roast upright by the bone, slice off the plump "lobe" of meat (the front of the thigh) by following along the bone with a long slicing knife. Now stand the roast on its cut surface and slice off the larger lobe of meat on the other side (the back of the thigh) by working your knife along the bone with a sawing action.

2 Remove the remaining meat from the bone so you have 3 chunks of boneless meat. Lay the chunks on their flat, cut sides, and carve thick slices downwards and against the grain, allowing 1 slice per person. The slices should be thick, almost like steaks. If they are too thin, they will be bloodless and the meat will be dry.

PORK & HAM

Cuts such as roast pork loin and shoulder, hams, tender fillet, saucissons, and salamis, with their many contrasting textures and flavours, are snipped, cut, sliced, and carved in different ways.

belly pork

Belly pork is a wonderful joint if prepared in the right way with a long, fine knife. On the bone, it provides the popular, Chinese-style spare ribs.

slicing **After boning, take the joint and square it off: cut it into a neat rectangle of even thickness. Trim away any excess fat. It is very easy to cut into slices of a similar thickness – simply press hard on the joint.**

slicing pork tenderloin

Be very gentle with this fillet: the meat is exceptionally sweet, tender, and delicate. Once the muscle sheath (connective tissue) is removed and the fillet is cut into portions, it cooks in minutes.

1 Slip a boning knife under the muscle sheath without piercing the fillet. Pull up the sheath and run the blade over the meat surface with a gentle sawing action, to separate the strip of tissue from the fillet.

2 To slice the fillet evenly into 70–80g (2½–3oz) portions, mark it out from the centre: first in half, then in quarters, then in eighths, before you begin to cut. Use a fine knife, such as a utility knife.

shaping medallions

Choose large pork fillets from the butcher to make these elegant medallions. Pork fillet, like beef fillet, is exceptionally tender and uniform in shape, so these are really easy to do.

1 Squeeze the fillet gently, to make it stand proud and firm. Use a utility knife to cut it into 7 equal portions about 4cm (1½in) long; they should look like little logs at this stage. Stand the pieces on their ends.

2 Wrap a piece loosely in the corner of a large piece of muslin. With the side of a santoku blade, hit and flatten, and squeeze it into a round with smoothing motions. Tighten the muslin as it takes shape.

3 Each piece of fillet should be formed into a perfect round, or medallion, ready for pan-frying. This technique can also can be used with beef fillet.

carving ham on an electric slicer

Use this machine for a boned joint of air-dried meat, such as Parma ham, Serrano ham, Braesola, saucisson, or large salamis – especially those with fennel seeds. Cut only enough slices to eat at any one time. When you want to use the slicer, cut away a collar of skin (and any fat) to a width of about 5cm (2in) with a serrated knife. If you cut away any more, the meat will dry out and you will have difficulty cutting and serving it later. An electric slicer is also wonderful for cooked octopus, peeled and cleaned melons, and all cold meats.

Serrano ham **Make sure that the machine is running smoothly and choose a setting to determine the thickness of the slices. Push the meat down hard in the carriage, turn on the machine, and run a few test slices to make sure that the thickness is as required.**

slicing ham terrine

A terrine should be wrapped tightly and chilled in a refrigerator for at least 24 hours and up to four days, before being removed to cut it into picture-perfect slices. A ham and parsley terrine is shown here.

even slices **With a sharp slicing knife, cut the terrine into 2.5cm (1in) slices with cling film still intact. Lay flat and peel off the film; slide on plates with a fish slice or palette knife.**

top technique

To line a terrine, lay out 3 rectangles of cling film on top of each other. Wet the terrine and line it with the film, so it overhangs. Add a filling; pull one long side of the film tightly over the top. Repeat on the other side. Fold up the short ends. Top with a foil-wrapped cardboard rectangle and wrap the whole terrine very tightly in at least 3 layers of cling film.

carving ham off the bone

Ham freshly carved off the bone has a sweet juiciness and is perfect for serving at parties. Follow the internal bone structure, and take off the meat in three lobes. Use three knives for the best results.

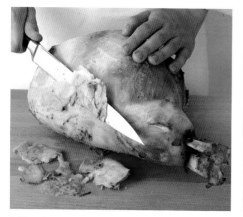

1 Use a long slicing knife to cut off the fat, then a flexible boning knife to remove the large lobe, from shank to the remainder of the ball and socket joint.

2 Follow the contours of the bone, which is now exposed, keeping your wrist flexible and twisting the blade, so that the blade strips off all the meat.

3 Take off the 2 remaining, medium and long lobes. Use the slicing knife to cut off the remaining sweetest and juiciest slivers that are close to the bone.

4 Place each lobe of ham with its flat side on the board and use a granton knife to carve the ham into thin slices. Serve them whole.

scoring fat for crackling

There are 4 simple techniques to get crisp crackling: score the rind, rub it with salt and oil, roast at a high temperature for the first 15 minutes, and do not baste the joint at all during roasting.

scoring **Cut across the rind widthways with a freshly honed paring knife, stanley knife, or a scalpel, keeping the lines parallel and close together. First work from the middle towards one edge, then turn the meat round and work from the middle towards the other edge. This is easier than scoring in a long line.**

carving roast pork

Transfer the meat to a carving board, cover loosely with foil and leave to rest in a warm place for 15–30 minutes.

1 Steadying the meat with a carving fork, slice between the crackling and the meat with a small slicing knife so the crackling lifts off in 1 piece.

2 With scissors, cut the crackling in half crossways to give short pieces that are easy to eat.

3 Carve the meat downwards and across the grain into thick slices, using a sawing action with a small slicing knife.

OFFAL

Offal should be eaten as soon as possible. The most famous dishes involving poultry livers derive from chicken, goose, or duck. All three are generally made into terrines or pâtés, but are equally good when fried as part of a salad, pasta, or rice dish.

preparing kidneys for frying

Supermarkets and butchers will prepare kidneys for frying, but they do not always do the job well. If you deal with them yourself, you can be sure of a good result. Whole veal kidney is shown here.

1 Carefully pull away and discard the white fat (suet) that surrounds the whole kidney – it will come away quite easily.

2 Lay the kidney upside down. With the point of a paring knife, cut around the fatty core and pull it away to release the membrane covering the kidney.

3 Discard the core. Peel the membrane off the whole kidney – it will slip off easily when you tug with your fingers.

4 Cut the kidney into bite-sized pieces following the natural lobes, then cut off the fatty cores from each piece. The kidney is now ready for frying.

preparing liver

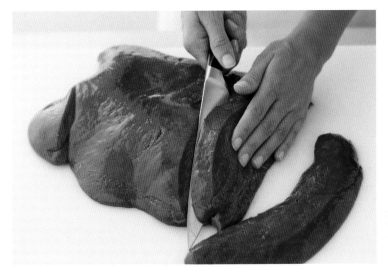

slicing **Calf's liver
is excellent for
frying because it
cooks very quickly
over a high heat.
For 6 people, buy
a piece weighing
800g (1¾lb).**

**Cut away any
membranes and
arteries with a
utility or small
slicing knife, then
slice the liver
thickly.**

cleaning chicken liver

trimming **Chicken
livers are always
good value. To
clean chicken
livers, use a paring
knife to cut away
any green patches,
membrane, and
the fibres around
the centre of each.**

CHICKEN

A fresh bird should have skin unblemished by dry patches, be plump relative to its size and weight, and should show no sign of bruising. Patches of dry skin indicate that the bird has been stored badly or frozen and it will need lots of extra butter or oil during cooking in compensation. Bruising is generally a problem with shot game birds. Red patches will turn dark and unsightly during cooking. The cutting method is identical for all poultry and the objective is to loosen the flesh from the bone with as little loss of flesh as possible. Focus on your own safety – use a regularly honed knife, keep the blade as close to the bone as you can, and never cut towards your hand.

parts of a bird

All poultry divides into two main parts, breast and legs. The winglets can be left attached to the breast or served separately as wished. Whether the breasts are halved or the legs further subdivided into thigh and drumstick will vary (see pp166–8), but this will be a matter of portioning, cooking time, or convenience rather than any major separation of differing types of meat.

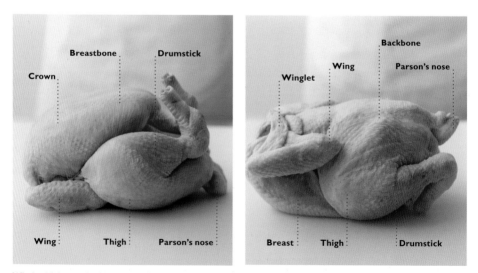

Whole chicken on its back

Whole chicken on its breast

removing the wishbone

Removing the wishbone is important if you want to make carving of the cooked bird or dissection into joints a great deal easier.

I Lay the bird on its back and lift the flap of skin from around the bird's neck. Run your finger around the neck cavity and you will feel the wishbone just in from the edge.

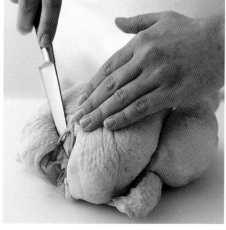

2 Using a freshly honed paring knife, scrape the flesh away from the wishbone so that it is exposed and clearly visible.

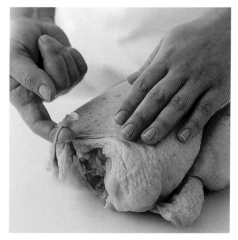

3 Run the blade of the knife just behind the bone, then use your fingers to lift and twist the wishbone free. Pull the skin back into place.

cutting a bird into four pieces

If you intend cooking the bird in any way other than whole, you will
need to know how to dismantle it into its component joints. All poultry
are formed in the same way, so the difference between jointing a
turkey and a partridge will be one of size rather than technique. Ducks
and geese are configured slightly differently, with long breasts and
comparatively short legs. This affects carving, but in essence they too
are taken apart in the same way, with similar knives.

1 Remove the wishbone (p165) from the bird, then
cut down through the skin between the leg and
the carcass, with a small slicing knife or utility knife.

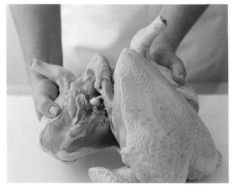

2 Bend the leg back as far as you can. The tip of the
leg bone, a ball and cup arrangement with the
backbone, will pop free.

3 Cut the leg away from the backbone, then repeat
the process with the other leg. Each leg may be
divided into thigh and drumstick.

4 Pull the wing out to its fullest extent, then use
poultry shears to cut off the winglet at the second
joint from the wing tip.

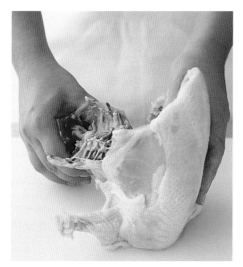

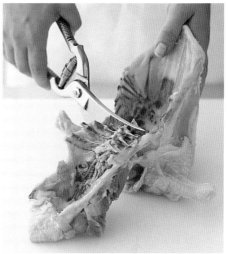

5 If a crown (the 2 breasts and wings on the bone) only is called for, snap the backbone of the bird at its halfway point.

6 Using poultry shears, remove the lower end of the backbone, which has no meat attached to it.

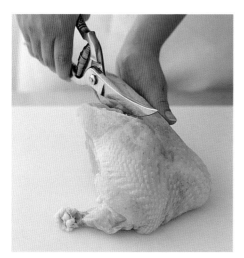

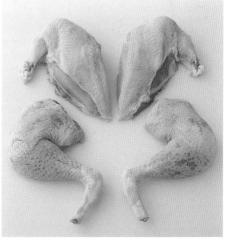

7 If the breasts are to be cooked on the bone, cut along the breastbone from neck to tail. Trim away any unwanted sections of backbone.

8 The chicken cut into 4 pieces. The legs take longer to cook than the breast, so for some recipes you will need to cook them separately.

cutting a bird into eight pieces

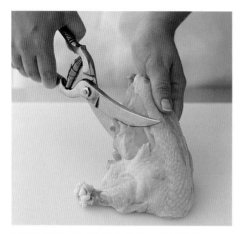

1 After cutting the bird into 4 pieces (pp166–7), use poultry shears to cut through the ribs two-thirds of the way along each breast and at an acute angle.

2 Each leg may be further divided. Locate the joint above the drumstick connecting it to the thigh and, with a small chef's knife, slice through to divide.

3 This is the chicken cut into 8 pieces. Many braising dishes call for the chicken to be divided into 4 or 8 pieces before cooking.

spatchcocking a bird

A spatchcocked bird has been flattened and transformed into
something more two-dimensional, which can be grilled evenly.
Poussins (baby chickens) are ideal for spatchcocking, and young guinea
fowl, quail, and squab pigeon can also be prepared in this way.

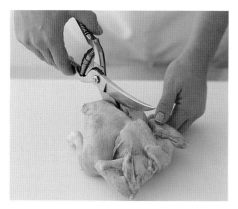

1 Turn the poussin upside down. Using poultry shears, cut along one side of the backbone, then cut along the other side and remove the backbone completely. Open out the bird and turn it over.

2 Using the heel of your hand or the flat side of a large chef's or santoku knife, lightly crush the bird all over. This tenderizes the meat so it cooks more evenly. With a freshly honed chef's knife, cut a few slashes into the legs and thighs for the same reason.

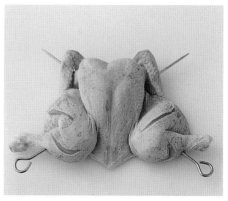

3 Push a skewer through the left leg to the right wing and another skewer through the right leg to the left wing.

4 If wished, the spatchcocked bird can now be marinated before being grilled or it can be roasted in the oven.

detaching the breast section from the bone

If the breasts are to be cooked as fillets, cut the breasts away from the bone. Work from the thick wing end, downwards. The breastbone will act as a natural brake, so you can press firmly either side of the central bone and follow its contours.

1 Using poultry shears, cut away the ribs and the backbone. Work from the thickest (wing) end of the breast towards the narrowest part.

2 Using a boning knife, separate the meat from the bone by following the contours of the breastbone, cutting the fillet neatly away.

3 Locate the small inner fillet on the underside of the chicken breast and slice any connecting membrane to remove it.

cutting a pocket into a breast fillet

for stuffing **The breadcrumbs of the stuffing should provide a good seal, but it helps if the stuffing is securely wrapped by the meat, so cut a pocket about 4cm (1½in) deep in the side of the breast fillet with a paring knife.**

boning a thigh

The thigh is easier to bone than the drumstick as the bone is straightforwardly positioned and visible. Keep the knife blade as close as possible to the bone itself and always cut away from your fingers. Also, angle the blade slightly towards the bone rather than away from it.

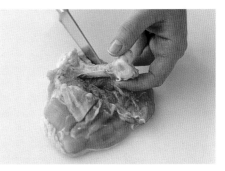

1 Place the thigh skin-side down on a chopping board. Use a small, sharp boning or paring knife to locate the bone at one end.

2 Cut through the flesh down the length of the bone. Draw the knife down the underside of the bone and cut the bone free.

boning a drumstick

The drumstick has a series of long, hard tendons along its length, which can complicate things but are best ignored rather than removed. Once the point of the knife is in contact with the bone, use a scraping motion to free the bone from the surrounding meat.

1 Starting from a point halfway along the drumstick, insert a small boning or paring knife through the flesh to locate the bone; slice along it to expose it fully.

2 Open out the flesh and cut carefully around the bone to free it from the flesh.

boning a leg

There are occasions when boning poultry is an advantage – for quick
braised dishes, perhaps – and occasions when it is essential, for
example when the meat is to be deep-fried or flattened into escalopes.

1 Place the leg skin-side down. Cut with a small chef's knife halfway through the flesh at the start of the thigh bone. Cut along the bone to the knuckle. Scrape to expose the bone, then ease it from the meat.

2 Perform the same operation from the knuckle down to the end of the drumstick. The bones will be exposed, but they will still be joined at the central knuckle joint.

3 Lift the bones up and away from the flesh and carefully use the knife to cut them free from the knuckle. A series of short nicks with the tip of the knife will do the trick.

chopping a whole raw bird

An alternative method to jointing a chicken (pp166–68) or duck is
to use a Chinese cleaver. A cleaver with a longer, narrower blade is
available for duck. It is easier to use a cleaver on a low work surface.

1 Hold the cleaver with both hands and line it up a fraction to the right of the highest point of the breastbone.

2 With dexterity and precision, bring down the cleaver to split cleanly through the bones. Push on the top of the cleaver once embedded in the carcass.

cutting cooked chicken with a cleaver

This method is quick and keeps the chicken hot. Instead of carving it off the bone, the Chinese cleaver chops through the bone, with the meat still attached, keeping it hot and juicy. This takes practice and can be dangerous, so be very careful. Choose a chicken with a rounded breast. Cooking it in Chinese style for several hours, possibly steaming it before roasting, softens the bones and makes it easy to cut this way.

1 Place the chicken, breast-side up, on the chopping board. Remove the legs in the normal way, through the soft cartilage of the joint, with a quick chop. Then chop off the wings from the breast, cutting through the soft cartilage of the joint.

2 Turn the chicken over. Bring the cleaver swiftly down close to one side of the backbone in one sweep, which will cleave it in half. Carefully chop out the spine from the other half, without damaging any of the breast beneath it.

3 Chop through each breast (while it is still on the bone) to produce even, thin slices.

4 Chop through each thigh, to cut it into even slices, but leave the drumsticks intact.

carving chicken

There are two aspects to carving poultry, one of which is better tackled by the traditional carving knife. This is the slicing of the breast into thin strips, especially on large birds such as turkey. The other is the division of the bird into joints and this is better done with a chef's knife.

1 Place the bird on its back on a chopping board. Hold it with a carving fork; use a carving knife to cut the skin between the leg and the breast. Next, draw the knife down and cut close to the breast.

2 Lift the leg backwards to release the bone from the body. This enables the cooked meat in crevices on the backbone to be taken off in one slice. Repeat the process for the other leg.

3 Hold the bird steady with the fork. Keeping the knife as close to the breastbone as possible, slice downwards and lengthways along one side of the bone to release the breast. Repeat on the other side.

4 You now have 2 legs and 2 breasts. To serve equal portions of both white and brown meat, divide these in half. Carve the breasts at a slight diagonal into equal pieces and put to one side.

5 Slice each leg through the joint to separate the thigh and drumstick. Place all the carved meat with the stuffing on a dish and serve.

skilful **carving** leaves the
least waste with very little left
on the **bone** or **carcass**

tools of the trade

A carving fork has a long stem leading
to short, curved prongs; when held
in reverse, these secure the meat
without damaging it. For serving, use
a fork shaped like a tuning fork with
two long, straight prongs and a short
stem, good for prodding and jabbing.

OTHER BIRDS

Ducks with their denser texture and elongated breasts need different handling to chicken. The point of carving however is simple – to render the meat easy to eat and presentable.

carving turkey legs

Separate the legs by cutting the skin between them and the main body. Press back each leg until it disconnects, then cut away the upper edge from the main body.

drumstick **To carve the drumstick, hold it upright by the bone and use a small slicing knife to cut the meat downwards into strips.**

thigh **For really large specimens, cut the bone out of the thigh with a small utility or paring knife. Slice down along the thigh to expose the bone, cut beneath it and remove. Slice the brown meat into strips.**

turkey breast

Keep the knife very close to the carcass and slice downwards and lengthways along one side of the breastbone to release the breast. The breast will come off in one piece.

carving **Lay the breast or a boned whole breast joint (shown here), flat-side down, and slice with a sharp slicing knife in the way you would cut a loaf of bread.**

rendering the fat from duck breasts

The sole drawback of duck for today's tastes is the fat content. There is plenty. All types of duck have a thick, rubbery membrane between the breast meat and the skin, making it impossible to remove.

removing fat **Take a freshly honed knife, eg a paring knife, Japanese knife (shown here), or scalpel. Slash the skin and fat, but not the flesh, in a tight crisscross pattern. Once the breast is cut, the fat will run out as it cooks.**

carving goose & Aylesbury duck

Goose and domesticated Aylesbury duck are carved more like chicken than wild duck. The meat will be cooked through and tender, so splitting the bird into sections rather than cutting into strips is needed.

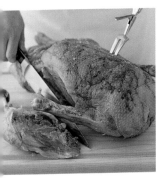

1 With the bird on its back, cut the skin between each leg and breast, using a long slicing knife. Lift the leg away from the body so that the thigh bone pops out, and cut it away at the base.

2 The leg can now be divided through the joint into thigh and drumstick. Keeping the carving knife as close as possible to the body, slice downwards and lengthways along one side of the breastbone.

3 Each breast will come off as a whole piece that may be subdivided into as many pieces as needed. The larger the piece, the longer it will retain heat. If desired, carve each breast on the diagonal into slices.

carving wild duck

The flesh of wild duck is denser and darker than other poultry and needs to be carved thinly. The secret of successfully carving a duck is to hold the breast in a vice-like grip so that it can be safely and easily sliced. Any strength and pressure should be on the hand holding the meat in place, not on the hand wielding the knife.

1 Keep the small slicing knife close to the breastbone at all times and slice downwards along the bone to remove each breast fillet.

2 Hold the breast firmly in place with a carving fork, then slice each breast lengthways into 4 strips. These can be fanned out or the breast can be reformed into its original shape.

quick tip

Wild duck is delicious if the breast is
served pink and the legs still juicy.
Turn up the oven when you take out
the duck. Carve off the legs through
the joints, located underneath. Wrap
in a layer of foil, with the pink sides of
legs facing upward. Place as high as
possible in the oven while you carve
the rest of the duck.

arrange the slices of
duck in a fan and
place the legs to the side

JOINTING A RABBIT

Start by placing the rabbit carcass so that it lies facing towards you with its back on the work surface, to get a good view of the cavity. The butcher will have removed most of the viscera.

1 Remove the kidneys and then snip out the liver with kitchen scissors; reserve. Turn over the rabbit, with its cavity downwards. To see where each muscly leg joins the loin, pinch the flesh at the top of the thigh.

2 With a short boning knife, cut in an arc through the ball and socket joint, towards the backbone. Run the blade along the backbone for the final part of the scooping cut; the leg should come away cleanly.

3 Cut off the other leg and turn the rabbit around so the exposed backbone points away from you. Lift each foreleg, slip the blade under the spade end of the shoulderblade and cut off close to the rib cage.

4 Remove the other foreleg. Use the heel of a large chef's knife to cut off the remains of the backbone: if needed, press on the spine of the knife to cut through the cartilage and bone in a single chop.

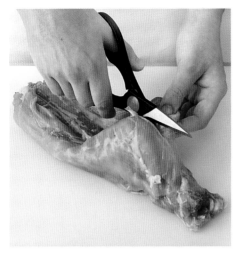

5 Lay the rabbit on its back and, with kitchen scissors, cut up through the breastbone, or sternum, so the breast meat falls apart in 2 flaps. Alternatively, use poultry shears, to cut easily through the cartilage.

6 Turn over the rabbit and tuck under the flaps. Cut through the rabbit, at the lower end of the rib cage, to leave just 4 rib bones attached to the loin. Square off the flaps, parallel to the torso.

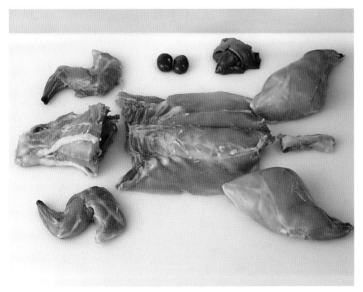

fully jointed rabbit
Wild rabbits prepared in this way may be stewed or casseroled; domestic rabbits may be braised. Use the bones for making a sauce or gravy.

The liver and kidneys may be chopped and fried in butter and served hot on pastry croûtes.

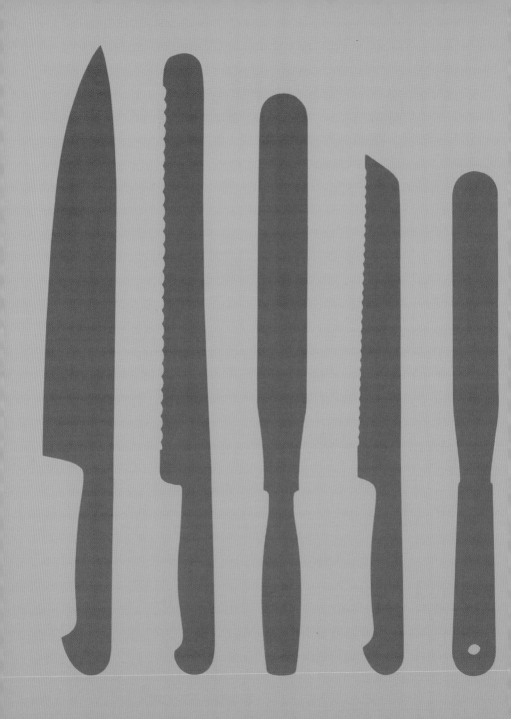

DOUGHS & DESSERTS

DOUGHS & DESSERTS

When a young chef decides to become a chef pâtissier, it's like taking holy orders. Cutting and slicing become a work of art; precision is the order of the day. Gâteau Pithiviers has hundreds of layers of puff pastry, scallops, cross-hatching, and a surface filled with the swirling glamour of radiating fine lines, all cut with the tip of a knife.

Don't let this deter you. If this type of cooking intrigues you, buy good-quality equipment from specialist suppliers, and be prepared to cut respectfully at all stages.

pastries & cakes

To avoid irredeemable mistakes, mark out shapes and lines with the spine of a knife before cutting items such as puff-pastry tarts and croissants. Never cut circles freehand; a set of round cutters will do the job perfectly.

The lighter and more delicious the pastry, as in pâte sucrée and pâte sablé, the more of a challenge it poses. The sugar and butter content of sweet pastries, choux puffs, and sponge cakes makes them delicate and friable after baking, so cut them with a granton knife, long scalloped knife, or electric knife for best results.

With all uncooked pastries, the freezer is your best friend; always keep space free for this purpose. You cannot cut oozing, limp pastries,

so just slide them on to an upturned, floured tray and place them in the freezer for 15 minutes or so. They will become firm enough to cut.

bread & pasta doughs

Bread dough can be soft and supple, tight and hard, puffy and slack – the simplest tools are needed for cutting and shaping it. With a little imagination and a pair of scissors, you can turn brown bread buns into appealing hedgehogs and breadsticks into wheatsheafs. Baked doughs have many kinds of crusts. To cut a loaf with a tough crust (like the sour-dough loaf shown opposite), you will need to use a knife with a serrated edge, to avoid wearing out your other knives.

Traditional pasta was always hand-cut: for noodles, fold the rolled dough into three, like a business letter and, with a large chef's knife, cut into pappardelle, and serve with ribbons (p60) of courgettes. Or use a ravioli cutter to cut squares of ravioli or rectangles for lasagne.

a final word

Even if the kitchen is filled with tantalizing aromas, it is important to allow breads and cakes to cool completely. This will release the steam before cutting, and enable you to achieve a perfect texture and slice.

RAW DOUGH

Although these doughs might look the same, each holds a secret in the making, with the revelation found only after baking. For example, the bread dough holds thousands of bubbles, and the puff and croissant pastry, dozens of layers. In the final shaping, cooking, and finishing, each dough needs to be cut or handled with the correct tool for a perfect result.

cutting out dough

A bulk fermentation of bread can be as large as a duvet, so you need to cut the dough into loaves or rolls fast. A dough scraper is not as sharp as a knife, so chopping hard on a work surface will not damage it.

quick tip

Maintain equipment with care: cutters should remain perfectly round; don't rest palette knives on a hot pan; and dust flour from all crevices before storing.

fermented dough **Fully fermented dough springs back when you press it with a finger and quadruples in size. Tip the dough on to the counter. If it is sticky, lightly flour your hands and the dough. Handle it gently, without compressing it. Chop into equal-sized chunks with the dough scraper – this cuts as sharply as a knife without damaging the work surface. Use for rolls, pizza bases, and flat breads.**

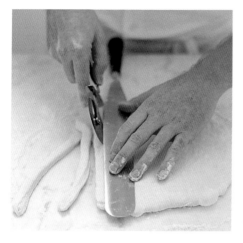

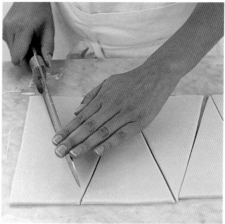

breadsticks **Pizza wheels are fun to use on dough, and make quick work of making breadsticks. As they cut at speed, they don't squeeze the air out of the risen dough. A long palette knife makes a good ruler.**

croissants **Place your fingers on the spine of a large, sharp chef's knife and bring it down in swift, straight lines so that you don't squeeze the dough. This will ensure that the buttery, crisp layers rise in the oven.**

puff pastry **The walls of a tart need to be straight to hold a filling. Place the knife point at the end of the pastry and cut with a clean, downward action along a palette-knife blade, so the layers rise without restraint.**

doughnuts **Position the cutter lightly on the dough with your fingertips. Bring down your palm quickly and hard to force it through the dough quickly. This cut will help the doughnuts to puff up in hot oil.**

working & finishing doughs

After you have taken the trouble to make a dough, you should take care over finishing it. Cutting fermenting doughs allows the dough to expand; cutting finished doughs creates a decorative effect. Using the appropriate knife helps to ensure that you achieve the best result.

slashing dough **To get the look and taste of breads such as baguettes, try cutting the surface of your dough with a razor-type blade, such as a scalpel.**

First mould the loaf, then when it is fully risen, draw the scalpel in lightning strokes across the dough, following the curve of the surface. Hold the blade tightly with your fingers, keep your forearm still, and with a firm but flexible wrist, draw the blade in an arc.

In the same way, the blade can be used for slashing fully fermented dough to check for signs of gas bubbles.

a decorative crust **To make a handsome brioche loaf, divide the dough into rolls and pack them into a tin. When risen and glazed, snip the surface of each roll in a cross with the points of sharp scissors.**

flipping a pancake **Using the rounded tip of a palette knife keeps your fingers away from the heat, supports the pancake without tearing it, and allows you to check that the base is cooked before flipping.**

BAKED DOUGHS

Baking a sponge or a batch of choux puffs and adding a rich creamy filling turns them into a fabulous dessert or tea-time treat. For a professional finish, discipline yourself to choose the right knife for the job, then cut neatly, sawing constantly and never pushing with the knife.

cutting & filling

To cut precisely delicate cakes, pastries, and fancies before adding a filling, use a long-bladed, serrated knife. It does not tear, and its length gives you control and finesse with tender textures.

delicate items **Once you have started cutting horizontally towards the hand holding the cake or pastry (profiteroles are shown here), transfer your grip to the top of the item. Alternatively, if it is fragile, steady it lower down, beneath the blade of the knife, to keep your fingers safe.**

I Score a guideline around the cake. By pressing on the knife using the thumb grip (*p37*), and sawing evenly and horizontally, you will easily follow the guideline, giving the layer an even thickness all round. Put your other hand lightly on top to steady the cake.

2 Gently lift the cut layer off the sponge and slip a palette knife or cake tin bottom underneath to transfer it carefully to a level surface. You could cut more layers, depending on the depth of the sponge. Once the filling is added, replace the layers.

Melba toast

Fragile, crisp, and brittle wafers of homemade Melba toast are one of life's little luxuries, to accompany a little caviar or pâté de foie gras. All you need is some sliced white bread and a very good bread knife.

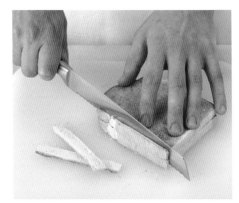

1 Toast the bread on both sides, under the grill or in a toaster, until it is pale gold. Pile 2 slices on top of each other. Cut off the crusts with a bread knife, trimming and cutting the slices so they are square.

2 Aim to complete the preparation while the bread is warm and flexible. Steady each slice with 1 hand and, with the blade parallel to the board, cut the bread horizontally through the middle.

3 Use a paring knife to scrape the untoasted bread off each slice and discard the scrapings. Cut each slice of toast into triangles. Place the triangles under a hot grill, untoasted sides up, and keep watch until they are pale golden and curled.

Remove from the grill, allow to cool, and keep in a dry place such as a warming drawer until serving.

cutting bread

Even a simple slice of bread can be elegant if cut with care and skill.
When slicing soft bread, use plenty of sawing motions, so that the knife
cuts cleanly and without tearing. It also helps to use very soft butter.

bread & butter **For the
thinnest slices of bread and butter,
take a slightly stale brown loaf and
cut off the end crust. Use a table
knife to spread the face of the loaf
with very soft butter, then cut off
a very thin slice with a bread knife.
(The butter helps to keep the slice
from tearing.) Repeat until you
have enough slices.**

**Serve with smoked salmon or
another smoked fish and a wedge
of lemon, with potted shrimps,
or with bread and butter pickle.**

sandwiches **Add your chosen
filling to buttered bread and close
the sandwiches. Pile 3 rounds on
top of each other. Grip a granton,
long serrated, or forged bread
knife with your thumb on top of
the bolster and cut off the crusts.
Trim the bread, as you cut, into
a perfect square.**

**Use the same grip to cut the
stack of sandwiches into triangles
(shown here) or into fingers. As
soon as the blade bites gently into
the first corner, place your hand on
top, pressing firmly and evenly to
hold the filling and bread in place.
Serve for afternoon tea.**

using an electric knife

The fast-moving, double blade is a superbly efficient cutting tool for challenging tasks such as slicing puff-pastry pies, champagne jelly, fruit terrines, and filled milles feuilles. Turn on the knife in a safe area, away from people and pets, and never touch the blade when it is plugged in.

pâté en croûte **This can take days to make and it is easy for an inexperienced chef to break the pastry, drag the jelly on to the filling, and ruin their work while trying to slice it – especially if the pie has very intricate layers.**

To use an electric knife, start the cutting blade and rest it on top of the pie. It will begin to cut through the pastry. There is no need to press it down: just let it drop through the pie, and the slice will fall away. The blade will not cut through the chopping board. Have a trowel spatula ready to catch the slice and transfer it to the plate.

cake slices **The electric knife will cut cleanly through a variety of textures such as nuts, chocolate, cream, fruit fillings, and sponge, keeping all intact and in shape without compressing any layers.**

Slice a cake as for the pâté en croûte (see above), without exerting any downward pressure, which could smear the ingredients. Simply let the blades ride gently through the cake for a perfect cut.

CHOCOLATE

The pedigree of chocolate is evident when you begin to chop it for melting. The best chocolate contains the highest amount of cocoa solids and cocoa butter (instead of vegetable oil) and is the hardest. You will recognize the brittle snap and sharp crunch when you cut it.

chopping chocolate **You need a serrated slicing knife and a plastic chopping board. Use a horizontal cutting grip (*p37*) and steady the tip of the blade with the palm of your hand. Chop the chocolate as finely as you can, into similarly sized pieces no larger than chocolate drops, so it will melt evenly.**

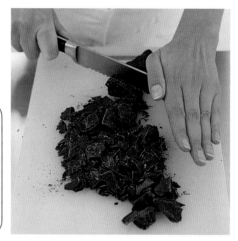

top technique

For easy shavings, stroke a chocolate block evenly with a vegetable peeler. Let the shavings fall into a small bowl and don't touch them – they will melt. Chill until required.

applying glaze **When the glaze is not too thick and not too thin, pour a shallow pool into the centre of the cake. With a flexible wrist, tilt a trowel spatula from side to side, so it pushes the chocolate towards the edge of the cake. The "legs" of the chocolate will form and drag the glaze evenly down the sides. Run the spatula vertically around the cake to finish it off.**

Don't be discouraged if your results are not like this the first time – this is truly a case of practice making perfect.

FRUIT

FRUIT

"The final course of a meal has to try the hardest: at least two courses have preceded it, so it needs to be exciting, brilliantly coloured, and gorgeous!" These are the words of Michel Roux, who won his Meilleur Ouvrier de France as a pâtissier, and indeed he is right. Thankfully, fruit, with its texture, juice, shapes, and its exquisite palette of colours, needs very little improvement and can make an impressive end to a well-cooked meal.

Cutting fruit for a fresh fruit salad is known as a "macédoine" after the kingdom of Alexander the Great, Macedonia – created by uniting the many and various lands that he had conquered. Alexander's territory is always depicted on historical maps as a collection of attractive shapes and colours.

attractive presentation

The breathtakingly lavish buffets served in some resort hotels are often the work of Asian chefs who practise from childhood how to make landscapes, complete with birds and dragons, out of fruits and nuts. These might be quite out of our reach, but we can improve our presentation with well-chosen garnishes that enhance the colour, composition, and overall design of the dish. They also release the fragrance of the fruit, especially citrus varieties, which are refreshing and enticing. With the correct cutting and garnishing tools, impressive results are easy to achieve.

Don't fall into the trap of overdoing it when preparing fruit. Too many twists and cuts can put people off because the food will look unappetizingly "over-handled", especially when it is raw. Keep the garnish as natural as possible: simply enhance the fruit's good looks, shape, and contribution to the dish. Choose the best fruits to work on; they should be ripe but firm. For example, a green mango, no matter how beautifully prepared, will be crisp and dry.

using fruit in drinks

As well as the dessert course, fruit makes a welcome addition to drinks. It makes all the difference to a jug of lemonade or a fruit punch on a hot summer's day. Iced glasses containing drinks garnished with fruits, such as orange segments, citrus zest, or melon balls, delight as much by their appearance as by their content. It is a bonus that these garnishes may be prepared well in advance, wrapped in cling film, and stored in the refrigerator.

peeling apples

There are several ways to peel and prepare apples, but if you want
shaped sections for decorative desserts, first peel with a swivel peeler.
Rub off any angular contours with a chef's knife, or peel them away,
so all is utterly smooth. Cut in half horizontally and core with a melon
baller (*see below*). Rubbing apples with a sliced lemon, or placing them
in water with lemon juice, will keep apples white while you work.

top technique

To make apple crisps, cut whole
green apples (skin, core, and all)
into very fine round wafers, on an
electric slicer or mandolin. Steep
in syrup and dry in a very low oven.

coring **Place the
melon baller on top of
the core, to introduce
the rim – the cutting
part – to the edge of
the core. Press hard
and scoop in one
motion: the core
should come out
as a single half-ball.
With a paring knife,
cut out any remaining
skin and the stem.**

preparing apple crescents

Golden Delicious apples are good cut into crescents for apple tart.
Peel the apples until they are smooth and all the same size. The better
shaped the apple at this stage, the more elegant will be the crescents.

I Examine each apple to establish at what angle
the core runs through it (it is not always perfectly
vertical) then, using an overhand grip, stamp out the
core with a corer. Remove the core from the corer.

2 Set the slicing blade of the mandolin to produce
slices that are no thinner than 3mm (⅛in). Run
each apple over the blade and allow the slices to fall,
leaving them in the order in which they were sliced.

3 Collect the slices and reassemble each apple on
your forefinger. Help the slices into a perfect
apple shape by shaking your finger gently. Stand the
re-formed apples at the edge of your work surface.

4 With a small chef's knife, cut each re-stacked
"apple" in half vertically to form perfect crescents.
Lay them in overlapping lines on a fine disc of puff
pastry, brush with butter, and bake in a very hot oven.

fluting apples for baking

Baked red apples in autumn are comforting, but the skins can be tough to eat. Removing the skin in strips, with a canelle knife or peeler, effectively half-peels them, leaving just enough skin for colour and to support the filling. It also allows the heat to penetrate to the centre.

I Cut channels vertically through the skin, with the canelle knife. To make a perfect pattern, cut the first channel, then one on the opposite side of the apple.

Cut the third channel halfway between the two existing channels, and cut the fourth opposite the third.

Cut the rest of the channels in the same way so that they are all equidistant.

2 Cut off the top of the apple, with a santoku or serrated knife. Hollow out the core with a melon baller (p198).

3 Spoon in a filling (here made with port-soaked currants, roasted and chopped hazelnuts, lemon, butter, and yoghurt) and bake for 15 mins at 160°C (320°F, gas 3) until the apples are soft and tender, but still holding their shape.

a little skilful cutting turns a homely dish into a stylish dessert

slicing a whole pear

This is an elegant way of presenting a pear for a dinner party. Use pears of the same size and shape and have enough sugar syrup with lemon juice ready in a saucepan to cover the pears, under a cartouche.

1 Core each pear through the base, using a melon baller; you will find the core runs only halfway up the pear. Retaining the stem, peel the pear with a swivel peeler, accentuating its beautiful curves.

2 With the tip and point of a serrated knife, cut a vertical line, all the way through to the middle of the pear, from stalk to base. Repeat every 1cm (½in) to create even parallel lines, without cutting the stem.

3 Carefully transfer the pear, with its stem still attached, to the plate. Lightly press the top with the palm of your hand, to splay it out. Use a turning knife to tease into position any misplaced slices. Spoon over a little syrup to make it shine.

making lemon-zest julienne

These are used in a wide range of recipes. One of the easiest and most useful options is to candy them in a boiling sugar syrup, to garnish many desserts. Choose fresh, large lemons with shiny, fragrant skins.

1 Peel the lemon using a swivel peeler, which is the most sensitive; remove only the zest, without the bitter pith. There is no need to square off (cut into a square or rectangle) the strips of peel.

2 If any pith remains on the zest, slice it off with a flexible knife, eg a filleting knife; keep the blade horizontal and parallel to the board. Use a chef's knife to cut the peel, with a rocking motion, into fine strips.

sectioning an orange

Who can resist a properly prepared orange? Serve with just salt, in salads, or for brunches. Choose fresh oranges with smooth, plump skins. Grapefruit and lemon segments are prepared in the same way.

1 Cut off the top and bottom with a scalloped slicer or freshly honed paring knife. Steady the orange with a fork and cut off the skin and white pith together, in sections; follow the curve of the orange precisely.

2 Cut with a sawing motion to the centre of the orange, keeping the blade flush with one side of a segment. Push off the segment, or cut it on its other side to free it. Repeat to cut out each segment.

cutting pineapple rings & chunks

When it is ripe, fresh pineapple is at its best and little of the chef's art is called for, but it does need to be properly peeled and cut. First, use a chef's knife to cut off the stalky top and then the base of the pineapple.

1   Stand the fruit upright and then slice off the skin in long strips. Cut from top to bottom and follow the contour of the fruit so that no flesh is lost where it bulges at the middle.

2   For rings, turn the fruit on its side and slice. Take a round cutter and cut out the centre of each slice. For pineapple chunks, cut the fruit into thicker slices and again remove the hard cores before segmenting.

fluted pineapple **First use a chef's knife to trim the base and peel the fruit (see above left), taking off all traces of the skin but leaving the "eyes". The spiral of the eyes will gradually become visible.**

Use the stalky top as a handle and start cutting from the middle. With a 15cm (6in) serrated knife, cut a V-shaped groove, just deep enough to remove all traces of the eyes. Follow the line of the eyes from the middle of the pineapple right around the fruit to the base. Turn the pineapple as you go.

Then cut a groove all the way to the top, and repeat until all the eyes have been removed.

this **impressively**
cut pineapple
may be used as a
decorative
centrepiece or
sliced into pretty,
fluted cartwheels

peeling peaches or nectarines

These are some of the most luscious fruits imaginable, full of sweet juice, and even more luxurious after skinning. Skinning is necessary for classical desserts, sauces, and purées. Make sure that you choose loose-skinned peaches and nectarines.

Cut a small cross in the skin at the base of the fruit with a paring knife.

2 Immerse the fruit in boiling water from a kettle for 30 seconds. Remove with a slotted spoon and transfer to a bowl of cold water. Remove from the water and pull the skin off with your fingers.

preparing a mango

Mangoes must be ripe unless they are an ingredient of a savoury stew or chutney. The stone is surrounded by hard fibres and the extent to which these affect the surrounding flesh varies. The best method is to make mango hedgehogs; it is unnecessary to peel them. Least is best.

1 Stand the fruit on its side; slice down and into the middle, with a utility knife. The fruit has a large, flat stone, so your knife must follow the soft flesh around it to the chopping board. Repeat the exercise with the other side, so the stone is cut off with a small slice of flesh; discard it.

2 Place the mango halves flesh-side up. Cut the flesh downwards into strips lengthways, then crossways until you reach the skin. Do not cut through the skin. Press the skin so that the fruit bursts upwards into segments. To remove the segments for a fruit salad, run a serrated knife close to the skin.

preparing melon boats

Don't be put off by the size and weight of melons. A scalloped blade
will make short work of the tough skin and opening up the melon
(a cantaloupe is shown here) to reveal soft, thirst-quenching flesh.

1 With a scalloped slicer, cut a small slice off the base
of the melon, so it stands steadily on the board. Cut
it in half and scoop out the seeds with a tablespoon.
Stand each half on its base and slice into wedges.

2 Run a flexible (eg filleting) knife along the inside
skin of each wedge, to release the flesh in 1 piece.
Push firmly on the skin and flex the blade flat against
the board so that it glides safely beneath your fingers.

3 Cut each slice into 5–6 even
pieces. Return to the skin,
staggering the pieces. Garnish the
melon boat with a citrus "sail".

quick tip

To make orange
"sails", use a canelle
knife to cut up to 12
channels into the skin
of a large orange.
Slice it thinly into
rounds, fold each
slice back on itself,
and secure into place
with a cocktail stick.

for a **dainty** and **refreshing** dessert, serve a selection of **melon balls** in a chilled glass

top technique

Halve or quarter various melons as needed and remove the seeds. Scoop out the flesh with a melon baller; swivel it 360° to create perfect orbs. Cut closely to get as many balls as possible.

apply this skill

Partly peel a pear to resemble a mountain snowline and use the poaching juice as a syrup. Peel kiwi fruits and oranges, but leave on edible skins, as of figs, to add colour.

Macédoine salad

The secret of a fruit salad is choosing perfectly ripe fruit, just firm enough to hold its shape, in as many colours as possible. Allow plenty of time to create neatly cut and pretty segments.

peeling & slicing **Use a peeler and a paring knife to clean fruits for a salad, cutting them into uniform and neat shapes such as crescents. Work on an immaculately clean surface and use a dough scraper to scrape all the juice from the board into the serving bowl.**

Drop apples, pears, and peaches into a light syrup or fruit juice straight after peeling, to stop them oxidizing and going brown. Add red and blue berries last; they can stain the syrup.

peeling and chopping chestnuts

Chestnuts come in several guises: raw, sweetened, or plain in cans or jars. Roast raw chestnuts by heating them on a griddle or open fire, or shell them after deep-frying, grilling, or blanching for a few minutes.

raw chestnuts **Whichever method you choose,** pierce the top of each chestnut with the point of a sharp knife to stop it exploding when hot. Grill or deep-fry for about 3 minutes until the shells split.

cooked chestnuts **Once they are cool enough to** handle, peel off the outer and inner skin. For stuffing, coarsely chop the chestnut flesh with a sharp knife.

now you have the fundamentals of a working relationship with good knives, you will find yourself taking **pride** in your cutting skills and being committed to excel.

GLOSSARY

acidulated Where acid is added, usually lemon juice (or white wine vinegar) to water.

balance Distribution of weight down the length of a knife, from handle to tip.

Beurre Nantaise Rich, "white butter" sauce, flavoured with finely chopped shallots and reduced white wine vinegar, originally served with fish in the Nantes area of France.

bevel (as on a single-bevel blade) Obliquely angled cutting edge, sharpened only from one side. Double-bevelled blades are also available.

bivalve Sea or freshwater mollusc that has a hinged, two-part shell.

blanch To dip food quickly into boiling water; to cook or pre-cook, depending on the ingredient.

brisket Meat from the breast of an animal.

bronze Alloy of copper and tin.

brown sauce Deep brown, glossy gravy, made with reduced meat stock, bones, and meat trimmings, and sweetened by a little chopped onion, carrot, celery, mushroom, and tomato.

brunoise Fine dice, eg of carrots, turnips, or courgettes, that are cubes of exactly 4mm ($\frac{1}{8}$in).

burr Rough ridge left on the cutting edge of a blade after sharpening.

butterfly To cut an ingredient, eg leg of lamb or large prawn, so it opens out like butterfly wings.

canon Loin of lamb, similar to loin or sirloin of beef.

carbon steel Alloy of steel and carbon; in knife blades, takes a very thin cutting edge that is easy to sharpen, but wears down quickly.

carriage Attachment for mandolin; protective casing that carries ingredient over the blades.

cartilage Firm but elastic tissue, blue-white in colour, found in the carcasses of young animals, and in the ends of the bones in mature animals.

cartouche Disc of greaseproof paper placed over braising, or poaching, ingredients, to keep them below the surface of the cooking liquid.

carve To cut cooked meat, poultry, or game into small portions or slices, from off the bone, using a slicing knife and meat fork.

carving knife A straight-edged, long, fine slicing knife: either 18cm (7in) or 26cm (10$\frac{1}{2}$in) long blade.

cèpe Strongly flavoured mushroom; edible boletus with a firm, bulbous stalk, highly prized in gastronomy.

cheek (as in knife blade) Thickness of the blade that tapers from the spine to the fine cutting edge.

chiffonade Thin, fine ribbons of a soft-leaved vegetable, eg lettuce.

chine Backbone; sometimes used as a verb eg to chine (cut away the backbone).

chuck Cut of beef from the forequarter, towards the neck, including part of the shoulder.

connective tissue Tissue that connects, surrounds, and supports muscles and organs in the carcass.

coral Colloquial name given to lobster eggs, which turn a coral hue, when cooked.

core (as in blade) Central, sharp, brittle blade, found in Japanese *kasumi* knives; it is supported by Damascene, or similar, steel in layers (some high-carbon, some low-carbon) to give the blade extra strength.

cos Type of lettuce; long and oval in shape with sweet, crisp leaves.

court-bouillon Clear, aromatic broth used mainly for the cooking of fish and shellfish.

crostini Crisp pieces of bread or savoury pastry, often with a savoury or sweet topping.

croûte Small, flat pastry case or toasted slice of bread; often with a savoury or sweet topping.

crustacean Mainly aquatic animal with a hard, close-fitting shell.

cutting edge Fine part of the knife blade that is honed, sharpened, and used for cutting.

daikon White radish that can grow up to 1m (36in) in length; also called a mooli and widely cultivated in the Far East. It is cut into fine strips, eaten raw with fish, or pickled.

Damascene effect The decorative patina on a polished knife blade, created by laminating the steel on to the core at differing angles; see also *kasuminagashi*.

deba hocho Slightly heavy Japanese knife, used for chopping and filleting; also known as a cleaver.

duxelle Basic preparation that consists of finely chopped mushrooms, onions, and shallots, softened in butter.

escalope Thin slice of white meat, usually from chicken or veal, but can be from large fish, eg salmon.

fillet Tender, boneless, long strip of meat or fish, or cut of beef taken from the backbone from the ribs to the loin – the most expensive part of the carcass.

forged knife Knife shaped from a single strip of metal by heating and hammering in a furnace.

French cut of best end Ribs of lamb with the rib fillet attached; also called *carré d'agneau*.

frisée Type of lettuce with a bitter flavour and frilly leaf; sometimes called chicory.

glaze Glossy icing, eg for a cake.

heft The weight of a knife, judged by lifting.

high-carbon stainless steel Blend of iron, carbon, chromium, and other metals, eg molybdenum, in a ratio making a stainless, resilient metal.

honyaki Japanese true-forged knives, made entirely of high-carbon steel.

kasumi Knife blade made from two materials: high-carbon steel (the *core*) and soft iron, forged together like Samurai swords. The steel provides the cutting edge and the iron forms the spine and cheeks of the blade.

kasuminagashi Meaning "floating mist"; decorative, shimmering effect on a laminated blade; *see also* Damascene effect.

laminated steel Steel strengthened and hardened by hammering it in layers parallel to the *core*.

ligament Short band of tough, flexible, and fibrous tissue binding two parts of the carcass together.

marinade Aromatic liquid, cooked or uncooked, in which ingredients are steeped; it enhances flavour, tenderizes, or prolongs the life of meat, fish, poultry, and seafood.

milles feuilles Sweet or savoury item of puff pastry, often with a creamy filling; the name,

"thousand leaves", refers to the flaky layers.

mirepoix Mixture of diced vegetables, including carrot, onion, and celery, usually used in the preparation of sauces.

muscle bundle Long, thin cells (fibres) bound together by sheets of connective tissue and organized in groups to form muscles.

pain bagna French baguette filled with ingredients including lettuce, tomatoes, olives, and olive oil; from the Nice region of France.

pâté en croûte Mix of spiced, minced, marinaded, and seasoned meats or poultry, baked in a pastry enriched with eggs. A meat stock is added through holes in the cooled crust and forms a jelly (aspic).

pâte sablée French sweet flan pastry, with a sandy, crisp texture; tends to be sticky and break easily.

pâte sucrée Sweet French flan pastry, fairly firm and made in a similar way to shortcrust pastry.

pinion Part of a bird's wing that includes the longest, strongest flight feathers.

roe Egg mass found in fish, or reproductive organs in scallops and other seafood.

roulade Savoury or sweet preparation, made in a similar way to a Swiss roll and then sliced.

saddle Backbone of lamb, between the ribs and the chump (pelvis); it contains the tiny fillet and loin (canon) on both sides.

sashimi Japanese delicacy, that consists of very fresh, raw fish and seafood, sliced into thin pieces about 2.5cm (1in) wide, 4cm (1¼in) long, and 5cm (2in) thick.

silk Soft, long fibres on sweetcorn cob; styles of a female flower.

shitake Mushroom with dark cap and strong, meaty flavour; traditionally grown in Japan. When dried,

known as Chinese mushrooms.

spatchcock To flatten a whole small bird or tender poultry by removing the backbone and pinning into shape with skewers, in order to grill or barbecue it.

stainless steel Alloy formed when chromium is added to carbon steel, to inhibit rusting.

sternum Breastbone; long, thin vertical bone that is attached to some of the ribs.

sushi Speciality of Japanese cuisine, consisting of small portions of vinegared rice, with thin slices of raw or cooked fish, and vegetables.

tako hiki "Octopus puller", a long, thin Japanese knife, used for slicing medium-sized fish and octopus, for sashimi. The blade can be pointed, but often is rectangular.

tomalley Fat or liver of the North American lobster; becomes green when cooked.

topside Top of hind leg of beef, above the shin; usually tough meat.

turn To make vegetable "rugby balls", about 5cm (2in) long, with seven faces; also known as *tourner*.

tunnel bone To remove the bone from a joint or carcass without piercing the skin, by creating a "tunnel".

vinaigrette Usually a cold sauce, made from olive oil, vinegar, salt, and pepper to which flavourings and aromatics can be added; used as a dressing for salads, fish, or white meat.

yanagi ba "Willow blade", a long, thin Japanese knife with a pointed blade; used for preparing medium-sized raw fish for sashimi.

zirconium oxide White, crystalline oxide of zirconium, that is used in ceramic knife blades. It does not corrode or stain, is extremely strong but brittle, and maintains a particularly sharp cutting edge.

RESOURCES

further reading

Beck, Simone, Bertholle, Louisette & Child, Julia, *Mastering the Art of French Cooking*, vol. I, Penguin Books, 1961

Botorff, Steve, *A Primer on Knife Sharpening*, Member of Ohio Knifemakers Association and Western Reserve Cutlery Association

The Culinary Institute of America, *The Professional Chef's Knife Kit*, John Wiley & Sons, 2000

Day, Christopher P., with Carlos, Brenda R., *Knife Skills for Chefs*, Pearson Prentice Hall, 2007

Ladenis, Nico, *My Gastronomy*, Ebury Press, 1987

Lane, Jenifer Harvey (ed.), *Larousse Gastromique*, Crown Publishers, 1998

McGee, Harold, *On Food and Cooking: The Science and Lore of the Kitchen*, Simon & Schuster, 1984

Norman, Jill (ed.), *The Cook's Book*, Dorling Kindersley, 2005

Parsons, Russ, *Difference between Western and Japanese Knives*, Los Angeles Times, 8 December 2004

Pauli, Eugen, *Classical Cooking the Modern Way*, CBI Publishing, 1979

Roux, Michel & Albert, *The Roux Brothers on Pâtisserie*, Macdonald, 1986

Yan-kit, So, *Yan-Kit's Classic Chinese Cookbook*, Dorling Kindersley, 1984

useful websites

en.wikipedia.org/wiki/Japanese_kitchen_knives
Good information on the history of knife-making, types of kitchen knife, and metallurgy.

usa.jahenckels.com
A German manufacturer; details on the history of Western knife-making, the importance of Solingen in the tradition, and the new Friodur steel-hardening process.

www.japanesechefknife.com/MASMOTO
Illustrated ranges of many Japanese knives; including kasumi and honyaki.

www.japanese.knife.com
Includes history of traditional Japanese knife-making, kasumi and honyaki knives, as well as advice on using and caring for knives, and interviews with professional chefs.

www.openlearn.open.ac.uk/mod/resource/view.php?id=198418
A case history of the kitchen knife, discussing the engineering principles involved in the development of the manufacturing processes.

www.tojiro.co.uk
A Japanese knife manufacturer's site; plenty of illustrated detail on traditional knife-making and Western-style Japanese knives, and Damascene laminated blades.

SAFETY & FIRST AID

Well-maintained and properly sharpened and honed knives may be a delight to use, but can also become very dangerous if handled carelessly. Even the best chefs have accidents from time to time, so be prepared to cope with any mishap. If you take a few simple precautions and learn some basic first aid, you should be able to minimize the impact of any injury.

Should you cut your finger or hand, rinse it under cold running water, and dry it thoroughly with a clean sheet of kitchen paper or clean cloth. Apply pressure on the wound to stop the blood flowing and apply a waterproof plaster. Hold your hand above your head. If the bleeding does not stop after five minutes of continuous pressure, seek professional help immediately.

Also seek urgent professional help if you see bone, plasma (thin, yellowish liquid) or anything other than clean skin and blood in the wound.

Cuts to the hand between the joints of the thumb and index finger require particular caution because there are larger blood vessels running close to the skin.

If you have recently cut your hand, wear plastic gloves while preparing or cooking food. If this is not practical, trim a finger from the glove, cover the wounded finger with it, and secure with elastoplast. Alternatively, use a proprietary finger cot (tubular bandage).

Always keep a first-aid box to hand in the kitchen: it should include plasters and dressings that are suitable for fingers and hands.

safety tips

• Make sure your magnetic rack is powerful enough, and your knives clean, or they can slip off. Stand knife blocks on an anti-slip surface.

• Keep knives sharp and honed and always use an appropriate knife for the task. Choose knife handles that you can hold in a safe, tireless grip.

• Sprinkle water on a dishcloth or kitchen towel, and place under the chopping board to stop it slipping or sliding about when in use.

• Cut away from your body and lay knives so they face away from you in the work area.

• Never run your finger along a cutting edge – use food, eg a tomato, to test its sharpness.

• Ensure that the work surface is tidy, to avoid knives being hidden. Never leave them in soapy water in the sink.

• Keep children well away from knives.

blade cover **Make a protective cover for any knife that has to be kept loose in a drawer. Fold some thick card to size around the blade and secure with sticky tape.**

INDEX

a

abalone: preparing 125
apples:
    baking 200–1
    crisps 198
    fluting for baking 200
    peeling 198
    preparing crescents 199
asparagus 90
    peelings: disposal 93
    preparing 93
aubergines 84
    cutting 54
avocados 84
    dicing 85
    halving and stoning 84
    peeling 85
    preventing discoloration 84

b

basil: slicing 94
batonnets 56, 72, 79
bean sprouts:
    preparing for stir–fry 82
beans:
    preparing 82
        for stir–fry 82
    removing strings 82
    topping and tailing 81
beans, green: preparation 80
beans, runner:
    cutting diamonds 60
    preparation 80
beef:
    cutting:
        across grain 142
        with grain 142
    entrecôte steaks: carving after
        grilling 143
    fillet steak: cutting for
        sautéing 143

beef (cont.):
    rib: carving 144–5
beetroot 66
    batonnets 72
    dicing 58
    julienne 72
beetroot leaves: trimming and
    slicing 75
bevels 20, 21
    sharpening 44
bivalves 101
blades:
    anatomy 12
    ceramic 21
    cleaning 38
    quality 23
    types 14
bolster 13
boning:
    chicken:
        breast section 170
        drumstick 171
        leg 172
        thigh 171
    fish:
        flat 117
        round 105–106
    lamb:
        leg 141
            butterflying 150–51
            tunnel–boning 152–3
        saddle 146–7
boning knife 25
box grater 33
bread:
    cutting 191
    Melba toast 190
bread dough: cutting 185
bread knife 26
breadsticks: cutting out 187
brisket 141

British knives 16
broccoli: preparing florets 77
Brutus grip 36
butcher's block 141
butter curler 34
butterflying:
    lamb 150–51
    prawns 126
butternut squash 78

c

cabbage:
    coring and shredding 76
    cutting 54
cabbage family 76–7
cakes:
    cutting 185
    slicing 192
calf's liver: preparing 163
canelle knife 34, 61
carbon steel 17
cardoons 90
carrots 66
    batonnets 56, 79
    cutting stars 61
    mirepoix 55
    preparing for stir–fry 82
    turning 73
carving:
    beef:
        entrecôte steaks after grilling 143
        roast rib 144–5
    chicken 174–5
    duck:
        Aylesbury 177
        wild 178
    goose 177
    ham:
        on electric slicer 158
        off the bone 159
    lamb: leg 155

carving (*cont.*):
  pork 160, 161
  tools for 141
  turkey 176
cauliflower: preparing florets 77
celeriac 66
celery 90
  mirepoix 55
  peeling 93
cèpes 89
ceramic blades 21
  sharpening 44
Chaucer, Geoffrey 16
cheese slicer 34
chef's knife 22, 25
chervil: chopping 95
chestnuts:
  chopping 211
  peeling 88, 211
chicken 164–75
  boning:
    breast section 170
    drumstick 171
    leg 172
    thigh 171
  breast:
    cutting pocket into 170
    detaching from bone 170
  carving 174–5
  chopping whole raw bird 172
  cooked: cutting with cleaver 173
  jointing:
    into eight pieces 168
    into four pieces 166–7
  meat 141
  parts of bird 164
  removing wishbone 165
  spatchcocking 169
chicken liver: cleaning 163
chiffonade 74
chillies 84
  chilli–flower garnishes 87
  preparing 87
    protection while 87
Chinese cleaver 17, 29
Chinese knives 17

chocolate:
  applying glaze 193
  chopping 193
  shavings 193
chopping boards 10
  for pungent ingredients 64
choux puffs: cutting and filling 189
citrus zester 34
clam knife 29
clams 101, 113
  buying 122
  opening 123
  purging 123
  remaining closed 122
cleaver 17, 29
cook's knife (chef's knife) 22, 25
coral: lobster 132, 133
corer 34
coriander: chopping 95
courgettes 78
  batonnets 79
  cutting stars 61
crabs 101, 122
  blue: cleaning 128
  cooked:
    removing meat from 130–31
  Dungeness:
    removing meat from 130–31
  soft–shell 101
    cleaning 129
croissants: cutting out 187
crustaceans 122
cucumbers 78
  making cups 79
cutler 16
cutters: set of 33
cutting:
  common mistakes 10
  correct ways of 10–11
  safety 38
  stance for 10
cutting tools:
  accessories 32–3
  early 16
  sharp–bladed 30–31
cuttlefish: preparing 136

d

Damascene effect 18
deba hocho 18
deveining: prawns 126
diamond stones 42
diamonds: cutting vegetables
  into 60
dicing:
  fish 112
  vegetables 58–9, 62, 85
double-handed grip 37
dough cutter 29
doughnuts: cutting out 187
doughs 185
  baked: cutting and filling 189
  decorative crust 188
  fermented: cutting out 186
  finishing 188
  raw: cutting out 186–7
  working 188
Dover sole:
  serving whole 121
  skinning and filleting 116
drawers: for knives 40
drinks: fruit in 197
Dublin Bay prawns *see* langoustines
duck:
  Aylesbury:
    carving 177
    cooking legs 179
  jointing 166
  rendering fat from breasts 177
  wild:
    carving 178
    meat 141

e

eel 113
  skinning and gutting 110
egg slicers 34
Egyptians 16
electric knife 24
  using 192
    for doughs and deserts 192
    for ham 158

f

fennel:
  bulb 90
  Florence 90
filleting:
  Dover sole 116
  flat fish 118–19
  round fish 107, 109
  skate wing 120
filleting knife 25
fish 101–21
  cooked 101
    flat 114–21
    boning 117
    filleting:
      four fillets 119
      two fillets 118
      gutting and trimming 114
    scaling 115
    serving whole 121
    skinning 115
  raw 101
    dicing 112
    fillets: serving 112–13
    slicing 112
  round 102–11
    boning:
      from back 106
      through stomach 105
    filleting 107
    gutting:
      through gills 103
      through stomach 102
    scaling and trimming 104
    scoring skin 107
    serving whole 111
    skinning fillet 108
fish scaler 29
Florence fennel 90
flounder: cutting two fillets 118
forging 21
French knives 17
fruit 197–209
  in drinks 197
  macédoine 197
  macédoine salad 211

fruit (cont.):
  peeling and slicing 211
  presentation 197
fruit vegetables 84–9
fungi: dried: soaking 89
fusion knives 20–1

g

game:
  anatomy of 141
  tools for cutting and carving 141
garlic:
  cloves:
    chopping 95
    preparing for roasting 64
  garlic paste 65
  peeling and chopping 65
  pungency 62
garnishing tools 34–5
Gâteau Pithiviers 185
general-purpose grip 37
German knives 17, 22
ginger: grating 97
globe artichokes 90
  preparing bottoms 91
  trimming to serve whole 90
goose:
  carving 177
  jointing 166
granton edges 21
granton knife 26
grapefruit: sectioning 203
graters 33
gravlax: slicing 109
greens, hearty: trimming and
  slicing 75
gremolata 65
grinding wheel 21
grips 36–7
guinea fowl: spatchcocking 169
gutting:
  eels 110
  flat fish 114
  round fish 102–3
  scissors for 103

h–i

ham:
  carving:
    on electric slicer 158
    off the bone 159
  terrine: slicing 158
hamachi 113
hearty greens: trimming and
  slicing 75
heft 21
Henckels, Peter 17
herbs 94–5
  rough chopping 94
  slicing 94
honbatsuki 18
honing 45
  freehand 47
  on steady steel 46
honyaki knives 18
horizontal cutting grip 37

j

Japanese knives 17–18, 20–21, 22
  sharpening 44
Jerusalem artichokes 66
julienne 55, 57, 62, 72, 203

k

kale: trimming and slicing 75
kasumi knives 18
kasuminagashi 18
kidneys: preparing for frying 162
kirenaga 18
kitchen scissors 30
knife block 40
knife–making:
  in Far East 17–18
  history 16–19
  in West 16–17
knives:
  anatomy 12–15
  blades:
    anatomy 12
    cleaning 38
    quality 23
    types 14

knives (*cont.*):
  bolster 13
  care 42–7
  choosing 9, 22–9
  cutting edge 12
  grips 36–7
  handles:
    anatomy 13
    ergonomic 15
    types 15
  handling safely 39, 217
  heel 13
  honing 45
  maintaining 9
  modern 20–21
  passing to others 39
  point 12
  quality: judging 22–3
  rivets 13
  safety 36–9, 217
  spine 12
  storing 40–41, 217
  tang 13, 15
  tip 12
  transporting 39
  types 24–9
  using 9
  walking with 39
  on work surface 38

l
lamb:
  for barbecues 149
  French cut of best end 148–9
  leg:
    boning 141
    butterflying 150–51
    carving 155
    studding with garlic
      and rosemary 154
    tunnel–boning 152–3
  marinating 149
  saddle: boning 146–7
laminating 21
langoustines:
  opening tail shells 127

leafy vegetables 74–7
leeks:
  cooking 62
  cutting diamonds 60
  cutting into julienne 63
  mirepoix 55
  washing 63
lemon grass: bruising 96
lemons:
  sectioning 203
  zest julienne 203
lettuce: chiffonade 74
liver: preparing 163
lobsters 101, 122
  cleaning 132
  cooked: taking meat from 134
  cooking 133
  eggs 133
  splitting 133

m
machine grinding 17
mackerel 113
mandolin 30
  using 57, 70
mandolin carriage 31
mangetouts:
  preparation 80
  removing strings 82
mango: preparing 207
marrow 78
meat 141–61
  anatomy of 141
  tools for cutting and carving 141
meat fork 33
Melba toast 190
melon baller 34
melons:
  preparing balls 209
  preparing boats 208
mezzaluna 30
  using 95
microplane grater 33
Miki City 18
mint: chopping 95
mirepoix 55

molluscs 122
molybdenum 21
monkfish: filleting tail 109
mushrooms:
  button: chopping 89
  dried: soaking 89
mussels 101
  remaining closed 122

n
nectarines: peeling 206
noodles: cutting 185
nutmeg: grating 97
nutmeg grater 33

o
octopus 113, 122
  cleaning and sectioning 137
offal 162–3
onion family 62–5
onions:
  cooking 62
  mirepoix 55
  peeling and dicing 62
oranges:
  making "sails" 208
  sectioning 203
oyster knife 29
oysters 101
  buying 122
  opening 122

p
palette knife 29
pancakes: flipping 188
Parmesan grater 33
Parmesan knife 29
parsnips 66
  batonnets 79
  cutting ribbons 60
partridge: jointing 166
pasta dough: cutting 185
pastries: cutting 185
pâté en croûte: slicing 192
pattypan 78
peaches: peeling 88, 206

pears: slicing whole 202
peas:
    preparing 82
    for stir–fry 82
peelers 34, 66
peppers: sweet 84
    preparing 86
pike: gutting through stomach 102
pineapple:
    cutting rings and chunks 204
    fluted 204–5
plaice: gutting and trimming 114
plums: peeling 88
pods 80–83
pommes allumettes 69
pommes gaufrettes 71
pommes pailles 69
pork:
    belly: slicing 156
    carving 161
    crackling:
        crisp 160
        cutting 160
    roast: carving 160
    scoring fat for crackling 160
    shaping medallions 157
    tenderloin: slicing 156
Portuguese 18
potatoes 66
    French fries 68
    lattice 71
    pommes allumettes 69
    pommes gaufrettes 71
    pommes pailles 69
    Pont Neuf fries 68
    turning 73
poultry 164–79
    anatomy of 141
    tools for cutting and carving 141
    see also chicken, etc
poultry shears 30
poussins: spatchcocking 169
prawns 101, 122
    butterflying 126
    deveining 126
    flattening 126

prawns (cont.):
    peeling 126
profiteroles: cutting and filling 189
puff pastry: cutting out 187
pumpkin 78

q
quail: spatchcocking 169

r
rabbit: jointing 180–81
radishes 66
    julienne 57
Réaumur,
    René Antoine Ferchault de 17
red mullet: filleting 107
ribbons: cutting vegetables into 60
root vegetables 66–73
    preparing 66
Roux, Michel 197

s
safety:
    cutting 38, 54, 217
    first aid 217
    grips 36–7
Sakai 18
salad leaves 74
salmon 101, 113
    cutting steaks 108
    scaling and trimming 104
    skinning fillet 108
    slicing gravlax 109
salsify 66
    batonnets 79
san mai 18
sandwiches: cutting 191
santoku knife 21, 26, 97
sashimi 101, 112–13
    fish for 113
sautéing: cutting fillet steak for 143
scaling fish 104, 115
scalloped knives: sharpening 44
scalloped slicer 26
scallops:
    opening 124

scallops (cont.):
    preparing 125
scalpel 29
scampi see langoustines
scissors 30
    for gutting fish 103
sea bass:
    boning:
        from back 106
        through stomach 105
sea urchins:
    cleaning 137
    roe 113
seafood:
    raw 101
    tools for 101
    toxicity 101
seeds 80–83
Seki City 18
serrated edges 21
serrated knives 26
    sharpening 44
sharpening 20, 21
    across stone 43
    along stone 44
    angle of 42
    general rules 42
Sheffield 16
shellfish 101, 122–37
shitake mushrooms 89
shoots and stalks family 90–93
shrimp deveiner 33
skate wing: filleting 120
skinning:
    Dover sole 116
    eels 110
    fish fillets 108
    flat fish 115
slicer: scalloped 26
slicing knife: straight–edged 26
smelting 16
Solingen 17
sorrel:
    sorrel sauce 74
    trimming and slicing 75
spatchcocking 169

spatula, trowel 29
spices 96–7
  bruising 96
  grating 97
spinach: cleaning 75
sponge cake: cutting and filling 189
spring greens: trimming and
  slicing 75
spring onions: preparing for
  stir–fry 82
squab pigeon: spatchcocking 169
squash family 78–9
squashes:
  cutting batonnets 79
  summer: turning 73
  winter 78
    halving, seeding & peeling 78
squid 101, 113, 122
  cleaning 135
  scoring mantle 136
stainless steel 17, 21
stars: cutting vegetables into 61
steels 21, 45
  honing on:
    freehand 47
    steady 46
stir–fry:
  preparing vegetables for 82
  sequence for 83
stones 20, 42
  sharpening across 43
  sharpening along 44
storage systems 40–41
straight–edged slicing knife 26
sugarsnap peas 80
sushi 101, 112–13
  fish for 113
swede 66
  dicing 58
sweetcorn 80
  cutting off kernels 80
  preparing for stir–fry 82
swipe–through sharpener 42
Swiss chard:
  trimming and slicing 75
Swiss knives 22

t

tang 13, 15
terrine:
  ham: slicing 158
  lining 158
thumb grip 37
toko hiki 101
tomalley 33, 132
tomatillos 84
tomatoes 84
  chopping 88
  peeling 88
  seeding 88
trout: gutting through gills 103
trowel spatula 29
tubers: preparing 66
tuna 101, 113
turbot:
  boning 117
  cutting four fillets 119
  skinning 115
turkey:
  carving:
    breast 176
    legs 176
  jointing 166
  meat 141
  turning 73
turning knife 26
turnip leaves: trimming and
  slicing 75
turnips 66
  turning 73

u

unagi 113
utility knife 25

v

vanilla seeds: extracting 97
veal kidney: preparing for
  frying 162
vegetables 52–97
  batonnets 56
  cutting 54–61
    diamonds 60

  for flavour 53
  ribbons 60
  safety 54
  stars 61
diamonds 60
dicing 58–9
fruit vegetables 84–9
leafy vegetables 74–7
mirepoix 55
pods and seeds 80–83
preparing 53
  for stir–fry 82
root vegetables 66–73
shoots and stalks 90–93
see also carrots; onions, etc

w

waterstones 20
Western knives 16–17, 21
whetstone 42
wishbone: removing 165
wok cooking: sequence for 83
work surface 10
  knives on 38

x–y

yanagi ba 101

z

zirconium oxide 21

ACKNOWLEDGMENTS

Publisher's acknowledgments

Dorling Kindersley would like to thank the following:

Project editor Annelise Evans for her help in producing this book; Dawn Bates for editorial assistance; and Gadi Farfour, Phil Gamble and Alison Shackleton for design assistance; Dorothy Frame for the index.

All new step-by-step photography
Martin Brigdale, William Reavell

New cutout photography
Gary Ombler

New photography Art Direction
Sue Storey

Model for new photography
Chef Richard Edwards

Food stylists for new photography
Belinda Altenroxel, Abigail Fawcett

For loan of props for new photography
With thanks to Dexam International, the exclusive distributor for Zwilling J. A. Henckels knives in the UK, for supplying all Henckels products used in the new photography for this book; Simon Kinder of Magimix UK Ltd for Magimix electric slicing machine; John Wells (Master Butcher); Martin Brigdale; Randi Evans; Abby Fawcett; Ronald Green

Picture researchers
Julia Harris-Voss, Jo Walton

Picture credits
The publisher would like to thank the following for their kind permission to reproduce their photographs:
(Key: a=above; b=below/bottom; c=centre; l=left; r=right; t=top)

Getty Images: 40l; James Carrier/StockFood Creative 212; Johner Images 41; Red Cover/Jake Fitzjones 40r; Photolibrary: Foodpix/Photolibrary Group 19; Stellar Sabatier: 16bl; StockFood.com: Drool Ltd, William Lingwood 1; courtesy of Zwilling J.A. Henckels AG: 17tl, 17tr

All other images © Dorling Kindersley

For further information see:
www.dkimages.com

Text copyright
(Key: t=top; b=below/bottom; c=centre; l=left)

Text copyright © 2008 Lyn Hall: 8–47, 53–4, 56–61, 64–9, 71, 74, 81, 82 (introduction t, text bl), 83, 86 (except caption bl), 87, 89, 92, 94–7, 101, 122 (caption), 126–7, 133, 136 (except tip box), 141–2, 148–9, 156–9, 172 (chopping a whole raw bird), 173, 180–1, 185–93, 197–202, 203 (making lemon zest julienne), 204 (fluted pineapple), 205, 208–210, 211 (Macedonian salad), 212, 214–217; tip boxes: 55, 70, 73, 75, 79, 82, 93, 103, 106, 107, 110, 122, 124, 134, 179.
Text copyright © 2008 Shaun Hill: 164–71, 172 (boning a whole leg), 174–9, 203 (sectioning an orange), 204 (cutting pineapple rings & chunks), 206–7, 211 (peeling & chopping chestnuts).
Text copyright © 2008 Charlie Trotter: 55 (except tip box), 62–3, 70 (except tip box), 72–3, 75–80, 82 (caption c), 84–5, 86 (caption bl), 88, 90–91, 93 (except tip box), 102, 103 (except tip box), 104–105, 106–107 (except tip boxes), 108–109, 110 (except tip box), 111–121, 122 (introductions t & c), 123, 124 (except tip box), 125, 128–32, 134 (except tip box), 135, 136 (tip box), 137.
Text copyright © 2008 Marcus Wareing: 143–7, 150–5, 158 (caption bl, tip box); 160–63.